To our

FAVOURITE

INTERNATIONAL

Economist.

MEMOIRS OF AN
UNREGULATED ECONOMIST

BOOKS IN THE ALFRED P. SLOAN FOUNDATION SERIES

THIS BOOK IS PUBLISHED AS PART OF AN ALFRED P. SLOAN
FOUNDATION PROGRAM

MEMOIRS OF AN UNREGULATED ECONOMIST

George J. Stigler

Basic Books, Inc., Publishers

NEW YORK

Library of Congress Cataloging-in-Publication Data

Stigler, George Joseph, 1911–
  Memoirs of an unregulated economist.

  "Published as part of an Alfred P. Sloan Foundation program"—P.
  Includes index.
  1. Stigler, George Joseph, 1911–   . 2. Economists—United States
—Biography. 3. Chicago school of economics. I. Alfred P. Sloan Foundation.
II. Title.
HB119.S75A3 1988   330'.092'4 [B]   88–47673
ISBN 0–465–04443–3

To Chick

CONTENTS

Contents

PREFACE TO
THE SERIES

THE ALFRED P. SLOAN FOUNDATION has for many years had an interest in encouraging public understanding of science. Science in this century has become a complex endeavor. Scientific statements may reflect many centuries of experimentation and theory, and are likely to be expressed in the language of advanced mathematics or in highly technical terms. As scientific knowledge expands, the goal of general public understanding of science becomes increasingly difficult to reach.

Yet an understanding of the scientific enterprise, as distinct from data, concepts, and theories, is certainly within the grasp of us all. It is an enterprise conducted by men and women who are stimulated by hopes and purposes that are universal, rewarded by occasional successes, and distressed by setbacks. Science is an enterprise with its own rules and customs, but an understanding of that enterprise is accessible, for it is quintessentially human. And an understanding of the enterprise inevitably brings with it insights into the nature of its products.

The Sloan Foundation expresses great appreciation to the advisory committee. Present members include the chairman, Simon Michael Bessie, Co-Publisher, Cornelia and Michael Bessie Books; Howard Hiatt, Professor, School of Medicine, Harvard University; Eric R. Kandel,

University Professor, Columbia University College of Physicians and Surgeons, and Senior Investigator, Howard Hughes Medical Institute; Daniel Kevles, Professor of History, California Institute of Technology; Robert Merton, University Professor Emeritus, Columbia University; Paul Samuelson, Institute Professor of Economics, Massachusetts Institute of Technology; Robert Sinsheimer, Chancellor Emeritus, University of California, Santa Cruz; Steven Weinberg, Professor of Physics, University of Texas at Austin; and Stephen White, former Vice-President of the Alfred P. Sloan Foundation. Previous members of the committee were Daniel McFadden, Professor of Economics, and Philip Morrison, Professor of Physics, both of the Massachusetts Institute of Technology; Mark Kac (deceased), formerly Professor of Mathematics, University of Southern California; and Frederick E. Terman (deceased), formerly Provost Emeritus, Stanford University. The Sloan Foundation has been represented by Arthur L. Singer, Jr., Stephen White, Eric Wanner, and Sandra Panem. The first publisher of the program, Harper & Row, was represented by Edward L. Burlingame and Sallie Coolidge. This volume is the sixth to be published by Basic Books, represented by Martin Kessler and Richard Liebmann-Smith.

—ALBERT REES
President
Alfred P. Sloan Foundation
August 1988

AUTHOR'S PREFACE

CHARLES BABBAGE, the colorful mathematician-economist-philosopher-gadfly of nineteenth-century England, was characteristically unconventional in his autobiography: he devoted a long chapter to injustices he had received and a rather short chapter to honors and compliments he had received. All good things are scarce, including approval by one's colleagues in any calling. Babbage was a dissenter from much of the prevailing opinion in the intellectual circle in which he moved, and so have I been. With much smaller claims than he to the esteem of colleagues, I have been treated generously by my fellow economists. There is no anger and little complaint in this account of a half century in American economics.

I wish to thank friends who have read and commented upon most or all of this autobiography in manuscript. Foremost is Claire Friedland, who as usual did much to mend my error-prone ways. Others were Rose and Milton Friedman, Charlene and Richard Posner, John Hause, and Virginia and Stephen Stigler.

GEORGE J. STIGLER

MEMOIRS OF AN
UNREGULATED ECONOMIST

PROLOGUE

Are Economists
Good People?

ECONOMISTS were first identified as a separate species toward the end of the eighteenth century. By 1790 the great English philosopher-politician Edmund Burke had issued his mournful prediction of the future of Europe: "But the age of chivalry is gone. That of sophisters, economists, and calculators, has succeeded; and the glory of Europe is extinguished forever."[1] Economists would be entitled to a certain respect, I suppose, for being a small group who destroyed a great civilization, but that achievement would earn no admiration or affection. But, of course, Burke was wrong. The nineteenth century in his land proved to be one of the golden ages of man, filled with economic and scientific and even moral progress. By

[1] Edmund Burke, *Reflections on the Revolution in France* (1790; reprinted 1959, N.Y.: Holt, Rinehart and Winston), 91.

a sophister no doubt was meant a captious or fallacious reasoner, and they are never up to any good, but neither economists nor the calculators caused much trouble, and they may have even helped a little to create that glorious century.

Nevertheless, the convention of denouncing economists had been established and was pursued with enthusiasm by men great and small. I shall not recount those denunciations, which were generally stronger on malice than wit, but ask instead, Why has it been fashionable to abuse economists (even granting the possibility that they may deserve it)? The main reason is easily named— economists have been the premier "pourers of cold water" on proposals for social improvement, to the despair of the reformers and philanthropists who support these proposals.

The practice of dashing high hopes received great notoriety at the hands of Thomas Robert Malthus, a nice, young clergyman fresh out of Cambridge University. His father, Daniel, was indeed a reformer, a disciple of Rousseau and the English anarchist William Godwin, who in his 1794 book *Political Justice* had blamed the troubles of the world on institutions such as marriage and property. Thomas Robert found himself arguing with his father against the view that man could reach "perfectibility" (or at least virtue and comfort) if only social institutions were corrected. He believed he had discovered an insurmountable obstacle to perfectibility—the "passion of the sexes." This passion would inevitably produce more children than the world could feed, because the power of human reproduction far exceeded the capacity of man and nature to provide subsistence. Convinced by his own argument, Malthus anonymously published an *Essay on Population* in 1798. He acquired immortality, and both he and economists acquired a durable reputation as cold-blooded critics of plans for man's improvement. His pessimism was the source for the characterization of economics as "the dismal science." The customary opposition by economists to what appeared to be benevolent policies continued even though Malthus's theory of population became increasingly more unpopular with economists themselves after 1825 (because it was wrong). My favorite example of the skeptical economic mind was a gifted Irish economist-lawyer named Mountifort Longfield.

It was the custom of well-to-do people in some areas of England to buy wheat in periods of acute scarcity and resell it to the poor at half the price they had paid. Clearly these people were acting out of pure altruism; in 1834 Longfield argued (in his *Lectures*) that they were also acting in pure or nearly pure futility. The reason was simple: Given the shortage of wheat until the next harvest, the only way the poor could be helped was by having the well-to-do eat less—economics respects the laws of arithmetic. Mere transfers of money would not create wheat before the new harvest and indeed, as Longfield ingeniously showed, they would only enrich the grain dealers. The poor would end up paying as much of their own money for wheat as if no one had sold it to them at half price, and only the initial holders of wheat would gain. This most benevolent of actions was the most incompetent of actions.

A large part of the public role of economists has been to pronounce the harsh verdict of economic logic: on reselling wheat at half price to the poor in Longfield's time and on setting minimum wages or maximum interest rates by statute in our time. Economists are messengers who so often bring bad news, and so earn the reputation of such messengers.

I do not consider this public role to be either unnecessary or unimportant. A society that does things that are inefficient or perverse in their effects ought to be told so. Doctors are obliged to warn against nostrums that do nothing to cure and may harm, and engineers are supposed to warn the legislature against perpetual motion machines. So it is with economists. The case was well stated by Benjamin Jowett, a famous classical scholar and Oxford don. He had occasion in 1861 to defend the economists against the complaints of Frances Power Cobbe, a formidable Victorian feminist best known as an antivivisectionist.

I write a line to thank you for the little pamphlet you have sent me, which I read and like very much.

There is no end of good that you may do by writing in that simple and touching style upon social questions.

But do not go to war with Political Economy.

(1) Because the Political Economists are a powerful and dangerous class;

(2) Because it is impossible for ladies and gentlemen to fill up the interstices of legislation if they run counter to the common motives of self-interest.

(3) (You won't agree to this.) Because Political Econ-

omists have really done more for the labouring classes by their advocacy of Free Trade, &c., than all the philanthropists put together.

I wish it was possible as a matter of taste to get rid of all philanthropic expressions, 'missions,' &c., which are distasteful to the educated. But I suppose they are necessary for the collection of money, and no doubt as a matter of taste there is a good deal that might be corrected in the Political Economists. The light of the feelings never teaches the best way of dealing with the world *en masse*, and the daylight never finds the way to the heart either of man or beast.

You see I want to have all the humanities combined with Political Economy. Perhaps it may be replied that such a combination is not possible in human nature. Excuse my speculations.[2]

It is difficult to refuse such praise. Would it not be churlish and even ignorant to deny that economists have increased measurably the understanding of the world within which we live, which is what scientists are supposed to do?

If one agrees that economists may be necessary, it does not require one to like them any more than one is required to like morticians. Economists are neither distinctively good nor bad, no more or less virtuous or brave or generous or faithful than the sum of mankind, and certainly no more modest. If they, and of course that means we, have any claim to the favorable regard of others, it is

[2] E. Abbott and L. Campbell, eds., *Letters of Benjamin Jowett, M.A., Master of Balliol College, Oxford* (New York: E.P. Dutton, 1899), 173.

because they are a little different. Economics is the only reasonably well developed social science (see what I said about modesty?) in that it has an extensive, operable, tested theoretical system. Living in economic science makes economists think a little differently than other people, and I hope these differences will be interesting.

CHAPTER 1

Getting Started

MOST ACADEMICIANS lead sedentary lives and I am no exception. Not only have I never led—nor cautiously followed—a cavalry charge, nor robbed a bank, but neither have I guided a president or even a mayor into new folly. Still, this book is about my travels through economics, so I shall tell something about myself, with appropriate brevity.

I was born in the small town of Renton near Seattle in 1911 and spent my first twenty years in Seattle, until I had finished college. Then I headed east to graduate school. My father Joseph had come to the United States from Bavaria at the turn of the century. Here he met and married my mother, Elizabeth Hungler, who had come in her late teens from Hungary, then part of Austria-Hungary. Both of their families were farmers. I was their only child.

My father had been a brewer until prohibition drove

that activity underground. Thereafter he tried a variety of jobs. He was a strong man, and became for a time a longshoreman. One of my earliest memories is that of my mother crying as my father left for a meeting of the long-shoremen's union, appropriately equipped with that supplement to *Robert's Rules of Order*—a blackjack. Soon afterward my father entered the real estate market. The Great Depression began early and hit hard in Seattle, where the shipyards had been closed down and Boeing Aircraft was still a tiny venture. My parents bought run-down places, fixed them up, and sold them. By the time I was sixteen, I had lived in sixteen different places in Seattle. But my family had a comfortable if nomadic existence, and my father acquired an astonishing knowledge of Seattle real estate.

Recently economists, particularly Barry Chiswick, have been studying immigrants in America. Chiswick finds that immigrants often have to start by working at poor paying jobs; they may be young and untrained for any occupation (as was true of my mother) or unable to speak English (as was true of both of my parents when they came to the United States). Immigrants are a select group, however: unusually healthy, energetic, and not afraid of risks. Try to imagine sending an eighteen-year-old girl off by herself nowadays to a foreign land, where she doesn't even know the language, and telling her to support herself. It wasn't quite that bad for my mother; she ended up in Pennsylvania in a community where only German was spoken. Chiswick finds that after a decade or two the European immigrants have risen to higher levels of earnings than the native population of comparable age, education, and experience, and I suspect that was true of my father too.

I was also a typical product of the melting pot. Until I was about three I spoke only German. Then, as I began to play with a wider range of children I refused to speak it any longer, and by the time I studied it in college (it was not taught in high school in retaliation for Kaiser Wilhelm) I was almost a cipher in German. I could still play pinochle in German, and understand my father's German, in which the percentage of English words rose rapidly over time. Correspondingly, the percentage of German words in his English sentences fell. The two paths probably crossed while I was still quite young.

I think I was less typical than other immigrants' children in another respect: I was free to spend my time as I wished. Both of my parents worked hard and long until about the time I entered high school (I was then twelve), and even thereafter I was left largely to my own devices. These devices smacked of a much watered-down Huckleberry Finn. For example, friends and I would pull an abandoned rowboat out of the bottom of Lake Union, cover the bottom with tar, and paddle around endlessly. Of course at that time we didn't know how to swim.

Seattle and the Pacific Northwest were a wonderful part of the world in which to grow up. I joined Troop 44 of the Boy Scouts, an extraordinarily enterprising troop. It soon got permission to build a log cabin in a wooded part of Sand Point, then a small Navy air base. Aviation was informal in the mid-1920s. One day we saw a small *land* plane settling down in Mud Lake, a small pond adjoining Lake Washington. The pilot was grumbling as he made his way to shore; he knew it was going to be a bad day when he hit a tree that morning. The troop played a game of "chicken" with the pilots. We would be swimming from our raft in Lake Washington and the pilots would swoop down close to scare us into jumping off the raft. They succeeded.

As the Sand Point base grew, the scout camp became untenable, but Troop 44 found a new home. In 1928, the Great Northern Railroad completed the longest tunnel in North America, one eight miles long with the town of Scenic at its western end in the Cascade Mountains. Previously the railroad had used a higher, three-mile tunnel ending at Tye. The new tunnel made the old tunnel obsolete, and we had the happy idea of asking the Great Northern to give Troop 44 the town of Tye—and they did.

Tye once had been called Wellington. In 1910 a snow slide had caught a train standing at Wellington and carried the train into the valley below. Ninety-six people were killed in that tragedy. The railroad had built snow sheds thereafter, but in that day one could still find the twisted steel of the earlier accident in the valley. The town of Tye was burned down just before the troop took possession, but we did get a fine railroad station. Tye did not have the convenience of Sand Point, which one could almost reach by streetcar; in fact Tye was about fifty miles from Seattle. Nevertheless for some years we used the town fairly regularly. By then my close friend Howard Wills and I had become assistant scoutmasters, serving with, rather than under, a warm, generous man named Harold Ware.

In summer the troop would climb around the Olympic

Mountains across Puget Sound and in the Cascade Mountains to the east. By the end of a four or five day hike I would be weary and footsore, often wet, and utterly disenchanted with the food we ate. I would swear to myself that *this* was the last time, but the next summer off we would go again.

This account implies correctly that I did not work during summers, chores aside. Finally, when I was about sixteen, Howard and I went to the Wenatchee Valley and thinned apples. Apples grow in clusters and it is necessary to thin them so that each apple will grow unstunted. I recall that we received forty cents per hour plus bed and board. Three weeks of this work constitute my lifetime record of physical labor for wages. I had never been idle. I played tennis (though not well) half the summer day during my teens, and I have been a house painter and handyman all my life. Still, that was an awfully short career in the blue-collar trades.

My father was walking proof that Bavarians are a very different breed from the stereotypical Prussian. He was highly individualistic, and not at all averse to risk taking (in fact, at one time he had been a compulsive gambler). One day he came home with a violin acquired at an auction and thrust it into my hands (I was about twelve, I suppose). I dutifully sawed away on it for a few days, producing sounds that must have been ghastly, until a friend of my father visited us and asked if I had ever heard of rosin. The violin soon went back to the auction house.

In 1932 the winds of political change were blowing as hard as man could recall, and the state of Washington, long a Republican stronghold, was in the center of the storm. Only one state office, the superintendent of education, did not have at least one Democratic candidate and my father proposed to file for it. I remember my horror: "Dad, you can't possibly do that. You don't even speak English well." Such is the snobbishness of youth. He didn't file, and the incumbent superintendent was the only Republican to win a state office in the landslide Democratic victory. I may have cost my father and the citizens of Washington a good deal of fun, and if so, I humbly apologize.

I rather believe, in retrospect, however, that my advice was good even if the reason I gave for it was not: after all my father wouldn't have had to teach English. He was an intelligent man but he was a novice in politics, in bureaucracy, and in educational administration. A large number of successful businessmen have gone on to high administrative posts in the national government, and many—I think most—have been less than distinguished successes in that new environment. They are surrounded and overpowered by informed and entrenched subordinates, they must deal with legislators who can be relentless in their demands, and almost everything in their agency that should be changed is untouchable. Educational administrators often have the additional virtue of sanctimoniousness. Of course there may be some self-interest in this analysis. Had my father won, I might have been drawn into his work.

Some years later my father did run for the Seattle city council, one of about thirty candidates for several vacancies. His campaign expenses were $2.75 for some cards announcing his candidacy, and he received several thousand votes on the basis of his German name. Shortly thereafter he dropped into a tobacco shop where he often bought cigars. He was greeted by the proprietor, "I'm very sorry that you didn't win; we voted for you." My father bristled, "That's a lie; I didn't get one vote in this precinct." So much for the secrecy of the ballot.

I was an insatiable and utterly indiscriminate reader, and a most casual student, until I got to the University of Washington. There I got lots of good grades, although entirely too many in undergraduate courses in business administration, which I assumed would prepare me for a life of trade. Businessmen were hardly lining up for my services when I graduated in 1931, however, so I obtained a fellowship at Northwestern University and a year later received a master's degree. As I left Seattle for Chicago (in a 1926 Jewett my father had given me), no thought of an academic career, and in fact no very definite idea of any career, was in my head. Neither of my parents was well-educated, and that meant that the college courses I had taken were not only self-chosen but uninfluenced by a family culture such as that, for example, in which my sons were to be inevitably immersed. That absence meant a lot of foolish reading, an overemphasis on taking "applied" courses like real estate principles (as if there were any), and avoidance of mathematics.

During the year at Northwestern University I met one stimulating teacher, Coleman Woodbury, who first stirred my interest in an academic life. Only gradually did economics begin to appear to me to be an intellectual discipline worthy of full-time study in its own right rather than as a plausible introduction to a life in business. The shift in my interest and goal was already evident when I returned to the University of Washington for another year and then went east again, never to return for any length of time to Seattle. My destination was the University of Chicago, and without knowing it, I was trying out for the major leagues.

I chose Chicago because my Washington teachers told me (correctly, it transpired) that Frank Knight and Jacob Viner were good economists. Also I thought that Chicago would give more attention to the individual student than, say, Columbia or Harvard. Columbia in particular was then most inept in dealing with potential students. It sent out a printed pamphlet in which the paragraphs pertaining to me were checked, e.g., "Your transcript must be received by May 15" and "You must make a deposit of $25 by June 10." Later on I was a professor at Columbia for eleven years, and so acquired much respect and affection for its consistently high quality economics department, but in 1933 the institution seemed remote and impersonal. Harvard responded to my inquiry with a personal letter from the secretary of the chairman of the economics department. The response from Chicago was from the department chairman himself, Harry Millis. I suspect I chose correctly with respect to personal attention.

I stayed at the University of Chicago from 1933 to 1936 and indeed discovered that its economics department provided an intense intellectual life. The teacher who greatly influenced me and most of my fellow students was Frank Knight. He was an irreverent critic of scholars and institutions, and it was indicative of both his skepticism of all authority and his intellectual vigor that sometimes he could not read *his* copies of important books: They were literally filled (even between the printed lines) with his written comments! He was already moving out of economics into the study of philosophy and religion, into the latter as a profound agnostic. A famous hypothetical figure in economics is the "economic man"—a perfectly rational individual who calculates precisely the costs and benefits of every action and then undertakes those actions whose benefits exceed their costs. Knight once observed, in the course of a lecture in the University of Chicago's Divinity School, that the economic man and the perfect Christian would have one thing in common —neither one would have any friends.

In some respects, a more improbable Moses, if Knight could ever forgive the analogy, could not be designed. There he was in his office, smoking cigarettes—with the aid of a toothpick—down to lengths, or possibly they should be called nonlengths, that threatened his moustache. There he was in the classroom, pursuing a new and monstrous fallacy in that week's *New Republic*, while the experienced students watched with unkind amusement the efforts of new students to take remotely orderly lecture notes. I clearly remember the occasion on which we were told to withdraw forthwith from economics if we did not understand the analysis of Ricardian rent theory about to be presented, and how, fifteen minutes later, Knight explained that he himself had not understood it until two years before. Sometimes Knight would appear in a colleague's classroom, expostulating at some argument explicitly labeled as nonsense. He was a lovable, indomitable, improbable man, but his powerful influence on his students did not derive from his eccentricities or his charm. A major source of his influence was the strength of his devotion to the pursuit of knowledge. Frank Knight transmitted, to a degree I have seldom seen equalled, a sense of unreserved commitment to "truth." This harsh mistress must be served even when the service was dangerous or painful. No authority was too august to challenge. Indeed, Knight would not have hesitated to tell Gabriel that his horn needed tuning. No contemporary passion was so powerful as to escape critical scrutiny and usually denunciation. The compromises of expediency were simply alien to the world of this scholar. Thus, it would be an absurd question to ask him to which political party he belonged, for neither possessed a scrap of him. Yet, as his arguments with Paul Douglas will reveal (in chapter 12), he could be ruthless in his support of friends and disciples.

This devotion to knowledge was exemplified and its message reinforced by Knight's way of life. He was not a consultant to great bodies or small, whether public or private; he did not ride the lecture circuit; he did not seek a place in the popular press. He conducted himself as if the pursuit of academic knowledge was a worthy full-time career for a first-class mind. This conduct was becoming less common in economics even at the time he was impressing it upon his students, but I suspect that dedication to scholarship is an essential ingredient of a great teacher; students are too intelligent to believe preaching that is not practiced. A corollary of Knight's scholarship was his unfailing suspicion of authority, which, if anything, he may have overtaught to some of us. Yet his unwillingness to bow to any authority except reason led to a special form of antiauthoritarianism; there was not the slightest element of condescension in his relations with students. He listened at least as carefully to a suggestion from one of us as to one from a famous scholar, and, in fact, it was sometimes downright embarrassing to be accorded the attentiveness with which he awaited our inadequate views.

I eventually wrote my doctoral dissertation, a history of economic theory in the last third of the nineteenth century, under Knight. He was extraordinarily helpful and kind, and I could not (and still cannot) understand why so few students wrote their theses under him. I was one of only three or four in all his years at Chicago, so far as I know. Milton Friedman, Homer Jones, Allen Wallis, and I, with important help from Lionel Robbins, arranged the publication of a splendid collection of Knight's essays, *The Ethics of Competition*, in 1935 on the occasion of Knight's fiftieth birthday.

The second major figure at the University of Chicago was Jacob Viner, a premier international trade specialist with a stupendous knowledge of the history of economics. A story is told of how a young scholar rushed up to Viner and said, "I've just discovered the first American mathematical economist!"

Viner replied, "You mean Charles Ellet?"

"How did you know?"

"He is the only undiscovered early American mathematical economist."

Whether this story is true or not, it captures both Viner's vast learning and his mischievous sense of humor. He was a stern disciplinarian in the classroom, and his example and instruction are the source of the still powerful tradition of the careful use of price theory at Chicago.

In an early session of his famous course on economic theory, Economics 301, in 1934, Viner asked one student what factors determined the elasticity of demand for a product. After a sensible beginning about the availability of substitute products, the student added "the conditions of supply." Viner turned red and said, "Mr. X, you do not belong in this class." My spine, and probably all of those in the class, began to tingle. Later when I took Viner's course in international trade theory, he forced me to take a letter grade. It was the only course I took at Chicago for a grade; commonly registered students took an "R." In any event, I received an A, so my academic record at Chicago could be summarized as either "one A" or "straight A." With only about thirty graduate students in the department, grades were not important.

It was years later before I fully appreciated some of Viner's great virtues. He was incomparably less dogmatic than Knight (or Henry Simons), and considering the inadequacy of the knowledge then (and now) of the sources and effects of public policies, that was the only responsible attitude to take. His study of intellectual history was similarly more detached than was Knight's. Knight began a famous article on David Ricardo's theory by listing seven aberrations in his work. (Ricardo was a premier economist in the period 1810–1823, and left a lasting imprint upon the style of economic thinking.) Viner, on the contrary, analyzed Ricardo perceptively and sympathetically, so a reader could *hear* Ricardo.

Henry Simons was the forerunner of what the public and much of the economics profession now take to be the central position of the Chicago School—a devotion to private (competitive) markets to organize the production and consumption of goods, with only limited economic functions for the state. In 1934 his pamphlet *A Positive Program for Laissez Faire* captivated many of us: It was trenchantly written and crowded with bold proposals of new economic policies. Completed in the trough of the deepest depression in modern times, the pamphlet had an urgency which never left his work: Western society was near the point of no return. Here are some characteristic passages:

> It seems nowise fantastic, indeed, to suggest that present developments point toward a historic era which will bear close resemblance at many points to the early Middle Ages. With the disappearance of free trade within national areas will come endless, destructive conflict among organized economic groups— which should suffice, without assistance from international wars, for the destruction of Western civilization and its institutional heritage.

> If, as seems possible, both capitalism and democracy are soon to be swept away forever by a resurgence of mercantilism, . . . then to commercial banking will belong the uncertain glory of having precipitated the transition to a new era.[1]

I shall return to Simons's views when I discuss the Chicago School of Economics in chapter 10. I came to know Simons more intimately than I knew either Knight or Viner, and I had introduced him to Marjorie Powell. Marjorie was a friend of Margaret Mack, the girl I was

[1] Henry C. Simons, "A Positive Program for Laissez Faire" (1934), reprinted in Henry C. Simons, *Economic Policy for a Free Society* (Chicago: Univ. of Chicago Press, 1948), 48, 55.

going with. Henry and Marjorie married after I left Chicago.

I also spent a good deal of time at Chicago with John U. Nef, ostensibly an economic historian but also a philosopher (and the owner of a marvelous Chagall collection). I spent much less time with other luminaries on the faculty, such as Henry Schultz, the econometrician, and Paul Douglas, who later became a United States senator.

The vigorous argument among the faculty suggests that there was no dominant school of thought at the time. Viner and Knight had a running argument over whether the costs of producing goods and services were ultimately psychological costs (such as the irksomeness of labor) or simply the foregone benefits of the alternative use one might have made of the resources (such as the enjoyment of leisure). I no longer understand how two brilliant men could argue so persistently over what is at least dangerously close to an argument over words: The irksomeness of labor can consist of the pleasurable activities it forces one to forego. Nor even now do I understand their method of communication; they used students to carry the argument from one classroom to the other. Legend had it that one student began his answer to one question in the examination for the Ph.D. in economic theory, "Part A: Answer for Professor Knight; Part B: Answer for Professor Viner." Legend also said he was not encouraged to remain at Chicago.

Similarly, Knight disagreed with his own disciple, Simons. Knight believed that political life is emotional and unreasoning, whether under democracy or dictatorship. Here is a parable from one of his unpublished speeches:

As for telling the truth in political matters—well there is a popular story of a small boy who told the truth. Not George and the cherry tree story, but the equally famous boy who made the simple observation that an emperor had no clothes on. Scientifically, there is one fault in that story; it is unfinished. I think the author was a kindly, sensitive soul, and hadn't the heart. In the story, as a story, it is of course a merit. But in a scientific lecture it should be finished, and it will only take a few sentences: That evening the people awoke to the realization that they had no emperor and the wise men were anxiously discussing what to do. You can't imagine a man as emperor after he had solemnly paraded the streets as his bare self, can you? The wise men couldn't agree, of course, and the next day there was a war. And in a year a prosperous, happy nation had been destroyed and a civilization reduced to barbarism. All because a child made an innocent remark about a plain matter of fact. And back of that, because an emperor was fool enough to let people see the human being inside an emperor's togs—which certainly everyone knew was there. Truth in society is like strychnine in the individual body, medicinal in special conditions and minute doses; otherwise and in general, a deadly poison.[2]

At least as important to me as the faculty were the remarkable students I met at Chicago. W. Allen Wallis arrived from Minnesota the same year that I did and we

[2] Frank H. Knight, "The Case for Communism," unpublished speech to the National Student League, November 2, 1932 (mimeo), in Warren Samuels, ed., *Research in the History of Economic Thought and Methodology*, Archival Supplement no. 2 (Greenwich, CT: JAI Press, in press).

began a lifelong friendship. We appropriated an empty office, filched some desks and chairs, and conducted postmortems of each teacher's performance. The office was decorated with signs, one of which said: "Mathematics has no symbols for confused ideas," an absurd statement, so it did not matter that it was incorrectly attributed to Fermat. Allen's critical, probing mind was obviously highly congenial. Allen was president of the University of Rochester from 1962 to 1970 and he is currently undersecretary of state for economic affairs. Our paths crossed many times since, and in 1958 he brought me back to Chicago.

Milton Friedman had come to Chicago in 1932. He was already a formidable intellectual figure; he not only could think very rigorously and originally, but was so *fast* in his mental processes. Many years later one of Milton's former students, Reuben Kessel, told me that for a long time he was puzzled by what had happened when he took a problem to Friedman: Friedman's discussion seemed soon to jump on to other issues. Only gradually did Kessel realize that Friedman had quickly moved on to a higher level of sophistication in the analysis of the problem. One episode illustrates Friedman's precocity. Milton, then a twenty-three-year-old graduate student, had been in bed for a day or two with a cold, and when he reappeared he had written a new paper. The paper presented a demonstration of error in a method by which Arthur Cecil Pigou, one of the most famous economists in the world, proposed to estimate demand elasticities from family budget data. Friedman's paper was published in the *Quarterly Journal of Economics,* even though Pigou failed to see and acknowledge his error. I have often told Milton that if only he would break a leg while skiing (a favorite sport of his), he would amount to something!

Kenneth Boulding came to Chicago from England as a Commonwealth Fellow. (Commonwealth Fellows usually spent some of their time, and perhaps more of their fellowship grants, speculating on the English pound.) He was both clever and so brave as to engage in disputes with Frank Knight about the nature of capital. Two other excellent young economists were Sune Carlson from Sweden and Robert Shone from Britain. And last I mention a brilliant college senior who was taking graduate courses in economics, Paul Samuelson. Paul has told the story that it was Allen Wallis and I who persuaded him to take advanced mathematics and become a mathematical economist. I doubt that we can make any claims for credit for his outstanding career. The entire student body was small but it was in extraordinary intellectual ferment.

The students and junior faculty, of whom Albert G. Hart was the most prominent, organized a seminar and invited a stream of visiting scholars (Oskar Lange and Fritz Machlup, among others) to give talks that led to remarkably stimulating discussions. When Chairman Millis asked us whether we would like to have the seminar listed as a course, we declined for fear that the faculty would seek to have a voice in its conduct. This suggests that we were confident, even arrogant, young graduate students, and of course we were. As an example of our mischief, one day we read in the *New York Times* of a quiz that was loaded with questions such as this: A square has four edges, and a cube has twelve edges; how many edges does a four-dimensional cube have? Henry Schultz trusted Allen Wallis, and distrusted me, so we went to him to judge which of us had the correct answer—Allen, with the wrong answer (twenty-four edges), or I, with the correct answer (thirty-two edges). Of course he chose Allen's answer.

One further episode is worth recounting so as to reveal my no doubt deserved reputation for brashness. Chester Wright taught American economic history from 4 to 6 each afternoon, with a ten-minute break at 5 P.M. for the Social Science tea. The course was competent but unexciting so I invariably attended only half of each session. One day Wright lugubriously announced that someone had written an anonymous note admonishing him for all too often running the class beyond 6 P.M. Every head in the class turned and looked at me. My reaction was one of chagrin; I had not written the note! The gentle Professor Wright bore no malice. When I took the preliminary examination in economic history, he put in one question on the economic development of the Pacific Northwest because he knew I had grown up there. Or, come to think of it, was it reproof? I didn't know the answer.

I am convinced that at least half of what one learns at a college or university is learned from fellow students. They live together and they argue among themselves with a vigor and candor that are inappropriate in discussions with faculty members, even tolerant ones. If one could attract good students without a good faculty, one could run a fine university very economically.

Chicago had a strong influence upon us. One lesson I learned, or possibly overlearned, was that of skepticism toward received beliefs and authoritative reputations. Knight in particular was prepared to dispute the Ten Commandments. I suspect that we heard the word "nonsense" too often. I certainly came away believing that the popular acceptance of an idea was little support for its validity. On the other hand, the school failed to immerse some of us in what, in retrospect, was a major, irreversible wave of the future—the systematic use of statistical data to estimate economic relationships and to test economic theories. There were indeed two important figures in that movement at Chicago. One was Paul Douglas, who pioneered the statistical study of labor markets and, even more importantly, the theory of production (the study of the relationship between the amounts of labor and capital used in a process and the amount of a commodity or service produced). The other was Henry Schultz, who was grinding out demand curves for oats, wheat, corn, and other crops (that is, finding the relationship between the size of a crop and the price it would fetch). The fault of my neglecting what is now called econometrics was no doubt my own. I had not taken any courses with Douglas (who was feuding with my teacher, Knight), and I was not attracted by Schultz's pretentious teaching style (as can be inferred from the tale of four dimensions), but I should have exploited his interest in the field. So I had to acquire my great respect for, and modest skill in, empirical work from fellow economists such as Friedman and Wallis, and from a long apprenticeship later at the National Bureau of Economic Research under Arthur F. Burns, Solomon Fabricant, and Geoffrey Moore.

I did acquire a strong interest in intellectual history at Chicago, as my doctoral dissertation on the history of economic theory attests. So, while I was failing to get the deep training in mathematical and statistical tasks, which would have better equipped me to participate in the increasingly more rigorous economic analyses of the discipline, I was acquiring some mastery of a branch of economics that was permanently declining in professional esteem. This is a modest complaint aimed at myself: I had failed to predict the direction of economic research for the next forty years, although the movement was already under way. My interest in intellectual history has continued to this day.

Graduate school has changed somewhat since those days. One noticeable change has been in tuition. I paid $300 a year when I enrolled at Chicago in 1933, a bargain even when converted to $2,800 in 1988 dollars. A more important difference was in the standard of living. For example, none of the students had an automobile. I recall that Milton Friedman more than a decade later was taken aback when he offered to drive several students to a seminar at Northwestern University; each declined because he was driving himself. Fellowship aid was very scarce; I received tuition for a few chores around the department. But I believe that in essentials little has changed. We lived as a cohesive little society, enveloped in economics the entire day through, and that is still characteristic of graduate life at a competitive intellectual center.

In 1936 I received an appointment as an assistant professor at Iowa State College (now University) in Ames. On December 26, I married Margaret Louise Mack, whom I had met at International House, where we both lived in Chicago. Margaret, known as Chick to everyone, grew up in Indiana, Pennsylvania, where her father was a lawyer. She had come to the University of Chicago after graduating from Mt. Holyoke College and taking a short stab at teaching. The night before our wedding her father took her aside and warned her that Hungarians had a penchant for beating their women. James W. Mack exaggerated his younger daughter's prospective plight—I am only one-quarter Hungarian. We had three sons (Stephen, David, and Joseph), of whom I am both suitably and justly proud. Today Stephen is a statistician at the University of Chicago, David is a lawyer in Connecticut, and Joseph is a businessman in Toronto. Chick died in August of 1970 at the cottage on the Canadian Muskoka Lakes where we had spent every summer since 1946. That cottage continues to be the center of our family life.

CHAPTER 2

University Life

THE UNIVERSITIES in which I have spent my professional life have ranged from good to superb, and I have observed a fair number of the remainder. They are as diverse in their goals and their achievements as the churches or the business enterprises of America. Indeed there is much diversity within each of them: the strongest departments usually contain at least one mediocrity; the strongest universities have at least one department of embarrassing quality; even the weakest institutions have seldom been able to avoid all strong appointments.

This diversity is no occasion for surprise; everything connected with people exhibits diversity. The greatest scientists, not excluding Newton, have written some drivel. I never appreciated the cliché that so often appears in a book review of a collection of essays by various authors, that "the essays are of uneven quality." As if any author ever wrote a book in which the various chapters were of uniform quality.

Universities cater to more highly specialized human beings than most other callings in life. If X is a great mathematician, he will be more or less silently endured even though he dresses like a hobo, has the table manners of a chimpanzee, and also achieves new depths of incomprehensibility in teaching. His great strength is highly prized; his many faults are tolerated. That is less true in most other callings because in the normal progression of his or her career, an entrepreneur or a cleric or a bureaucrat or a lawyer must display a larger variety of talents to succeed. The business executive will have to learn something of finance and public relations as well as marketing or production; the successful cleric must be able to adjust comfortably to a richer and more sophisticated congregation; the bureaucrat must learn how to handle the dynamite called "congressmen"; the lawyer must please clients and courts.

I should not exaggerate the virtuous single-mindedness of the university's search for scholarly ability. Because of personal characteristics or behavior a maverick type like Thorstein Veblen or Abba Lerner found it much harder to receive the recognition his scientific abilities deserved. Also anti-Semitism was widespread in even major American (and European) universities before World War II. Members of universities, like so many other people, behave less admirably than we should.

Academic life has many fewer levels or types of activity than almost any other field. A senior professor is still doing mostly the same things he did as a young instructor. I am fond of illustrating that fact with an anecdote about my mother. When I was about thirty-four she asked me what position I had (itself a comment on her knowledge of academic life), and I proudly said, "A full professorship." A decade later she repeated the question and I repeated the answer, this time as a matter of fact. "No promotion," she remarked. I used to tease dear Elsie, but of course she was right; unless I went into academic administration or left academic life, my work would remain basically unchanged.

It is reasonable enough that the capacity to produce new knowledge should be prized most highly in the industry dedicated to the production of knowledge. It also seems reasonable to me that extreme specialization should be common in that industry. I believe the lesson that Adam Smith taught about how eighteen people who each specialize in making one part of a pin ("one man draws out the wire, another straights it, a third cuts it . . .") could produce 240 times as many pins per day as if each made the entire pin. Pins may not seem much like theorems, theories, experimental techniques, literary criticism, or historical syntheses, but all must be produced by people with finite capacities to learn and work. One can plow a deeper furrow in a smaller field.

Specialization, however, has more foes than friends. The liberal education most colleges still strive to instill in students is one manifestation of the drive for broadly educated people, and the heroic stature assigned to the Renaissance man is another. In a rich society such as ours, with the average length of life we now enjoy, we can surely afford to acquire skills and knowledge which will be useful to us primarily in our nonmarket activities.

Adam Smith in fact severely condemned the narrowness of vision and thought that he believed would result from excessive specialization. But was he being consistent? Shouldn't we also specialize in our liberal education? Shouldn't we learn to play a musical instrument rather than study music appreciation? Or should I keep to economics rather than discuss the content of the best liberal education? Like all good principles, specialization is troublesome when carried to the limit.

The crotchets and foibles of scholars should not obscure the strenuous competition that pervades the scholarly disciplines. The competition is seldom so personal and sharply focused as a footrace because two scholars at one school seldom are working on precisely the same problems. An active scholar is trying to change the work that is being done in his field, by changing the problems or the theories or the methods of dealing with the problems being studied by all the other scholars working in his area.

The competition may arise in the pursuit of the same goal—the famous rivalry for the discovery of the double helix between James Watson and Francis Crick and Linus Pauling has been candidly described in Watson's book, *The Double Helix*. Much more often the rivalry is indirect. Milton Friedman dominated the work in macroeconomics between 1960 and 1975, although he was outside of and hostile toward the dominant Keynesian approach. His attacks on the Keynesian system and his claims for a monetary approach to explain the level of the money income of an economy were the center of controversy among economists. I used to say that he controlled the Cambridge universities and Yale. They were devoting much of their efforts to seeking to refute what he had most recently written. As is customary in science, he did not win a full victory, in part because research was redirected along different lines by the theory of rational expectations, a newer approach developed by Robert Lucas, also at the University of Chicago.

The volume of the published output of the main competitors in a science is usually awe-inspiring. The champion in economics was Harry Johnson of the University of Chicago, whose bibliography runs to 52 closely printed pages, including about 500 journal pieces. Yet Harry died at the age of fifty-four, after a life as prodigious in other dimensions as it was in scholarship. Some of the other leaders are almost as prolific. Most but by no means all of Paul Samuelson's articles are combined in five fat volumes, and Friedman's journal articles number at least 150.

I remember once asking myself: How could Harry Johnson possibly have hundreds of ideas, or even five really fundamental ones? But that led me to ask the same thing about the eighty articles I had then published, with the same obvious answer. The vast output of prolific scholars contains much repetition and crossing of t's, but it also has a role to play in the important work of these scholars. The repetition, elaboration and application, and criticism of rival views, serve to educate the profession (and the scholar himself) to the ideas. The importance that the ideas acquire in turn leads to many invitations to speak and write about them. In any event, it is necessary to have energy as well as intellectual capacity to become an influential scholar. Perhaps professors should be tested to discover whether they are taking steroids.

An important way, if not the most important way, in which one influences a field is through one's students. They study the views of their major professors thoroughly—they will be examined on them—and sympathetically. The sympathy is important; it means that the student is looking for what is true and for ways to remedy what is incomplete or wrong. Able students will take over much of the task of persuading the profession of the importance of their professor's views. Only in the more academic of governmental bureaus like the Federal Reserve or the research centers of a few businesses, notably Bell Laboratories, is there likely to be comparable support for the work of a staff economist.

The better students, moreover, are at the better schools. Most important scholars I have known received their training at major graduate schools, no matter how ordinary their undergraduate education. (It is remarkable how often a famous professor, who as an undergraduate attended the closest state university, dies a thousand deaths if his children are not admitted to their first choice in the Ivy League.) On the other hand, many extremely promising undergraduates must go to the lesser graduate centers, because of personal circumstances, ignorance of their quality, better financial aid, or whatever. Why have so few of these latter students succeeded in research? After all, the material that is taught in the major graduate centers is in print and available everywhere. My explanation is that in the leading graduate centers the students learn primarily from one another. They learn to impose higher standards upon themselves, both in the selection of problems to work on and in the adequacy of the solutions they provide to these problems. Bull sessions are a more effective method of teaching and learning than classroom lectures or discussion. One colleague has said that he considered his role in the classroom to be that of providing topics for bull sessions. I don't think that the successes achieved by the graduates of the major schools are due simply to an old-boy network. Anyone who does outstanding work is in strong demand even if, like the hypothetical mathematician I referred to, he has few other redeeming traits.

Association with a group of able colleagues is a strong advantage that a professor usually has over a nonacademic economist. Frequent exchanges with strong minds and powerful scientific imaginations that have a deep understanding of the problems one is struggling with are invaluable in discovering errors and eliminating strange perspectives that creep into one's work. No academic economist could persist in the kind of illusion that captivated a magnificent German economist-farmer, Heinrich von Thünen (1783–1850)—that the just wage of a worker was the square root of the product of his subsistence requirements and his productive contribution.

One benefit of associating daily with another person is that communication of ideas becomes much more efficient. Even though Jones and I have always spoken English, and may even have gone to the same graduate school, each of us thinks somewhat differently; we each have a different order in which we think and probably a different pace in expressing ideas. Family members use words that have special meanings for them. A reference to "Z" brings to mind a tedious bore or a remarkable procrastinator; in our family, to "Lizzie Bean" was to lead out immediately all one's aces in bridge. So it is with every person, and that is why intimate association makes communication between people efficient and accurate. If I had known David Ricardo, I would be better able to understand his written words. That would be a help, because to this day the meanings of his theories are much debated.

Collaboration among economists has taken on a more literal role than mutually beneficial discussion in the past twenty-five years. If one looks at a volume of an economic journal (such as the *Journal of Political Economy*, of which I have long been an editor) published in 1900 or even in 1940, almost every article had a single author. Today, however, more than half the articles have two or more authors, who are by no means always located at the same school.

Collaboration has become so prevalent for a variety of reasons. The collaborators are often people of similar talents. Each author *could* have written the article alone, and indeed, if they are not of roughly equal abilities, the more able is making a gift to the less able. One obvious reason for and result of the collaboration is a much expanded list of publications for each author. Apparently writing half of ten articles does more for one's reputation than writing five articles alone. Again, because different minds have different ways of approaching a problem, the several authors stimulate each other, so in a way, joint authorship is the price each author pays to have the others think hard about his problem. One effect of the practice is troublesome: It is becoming difficult to estimate the quality of young scholars who seldom write alone.

For various reasons—able colleagues and collaborators, the support of students, the availability of time for research—and because the leisured gentleman has almost disappeared from our society, economics became almost exclusively an academic science by the beginning of the twentieth century. Today there are more economists in business and government than in colleges and universities, but the center of research is still in academia.

It was not always so. I have mentioned David Ricardo. He was a highly successful stock (government bond) broker during the Napoleonic wars who became interested in economics and became one of our discipline's most illustrious figures in eight short years, 1810–1817. John Stuart Mill, the ruling sovereign of English economics in the middle of the nineteenth century, spent most of his life as a senior "clerk" (executive) in the East India Company.

There are advantages to living outside a university (did Ricardo have to attend numerous committee meetings?), but they are clearly swamped by the advantages of academic life when it comes to doing economic research.

My Academic Itinerary

I started my professional career at Iowa State College in 1936. It was one of only two available academic posts (the other was Ohio State) known to my professors at Chicago in that year, and it would not have been available if Homer Jones had not turned it down. Homer was a clear-headed, strong-minded economist, who after a long career in Washington, D.C., eventually founded an important center of monetary economics at the Federal Reserve Bank of St. Louis. I later told him that if he had accepted the Iowa post, I probably would have become a Seattle real estate dealer. Anyway, it was a sumptuous position: $3,300 plus the privilege of teaching twelve hours a week. That teaching load now sounds onerous to a professor but no doubt positively luxurious to a non-academic; my father-in-law said I had a license to steal.

I can still vividly recall my very first class session, which was in "economic principles," then and no doubt now a collection of glimpses of economic theory and current economic problems. I had carefully prepared by outlining the first five or six weeks of the twelve-week quarter. About forty-five minutes into the class hour I found myself at the end of my notes! I was filled with consternation. I might last out the first session, but what about the rest of the quarter? I believe that this is not an unusual experience for new teachers, but I must admit that I have never reached the abundance of knowledge that made the time in the classroom seem inadequate.

My chairman at Ames was Theodore Schultz, who later went on to Chicago before I returned. He has an extraordinary intuition for important problems and has been a pioneer not only in applying modern analysis to agricultural problems but also in the study of economic development and the economics of education (for which he received the Nobel Prize). I had excellent students at Ames, and one, D. Gale Johnson, later was my chairman and then provost at Chicago.

In 1938, at a spring meeting of economists in Des Moines I met Frederic B. Garver of the University of Minnesota, and that meeting led to an invitation to come to Minnesota, which I accepted. Although I remained there until 1946, I was on leave for three years during the war, first at the National Bureau of Economic Research and then at an early operations research group called the Statistical Research Group at Columbia University.

Economics at Minnesota during my tenure was not in its present state of high prosperity, but it was a good department with able colleagues and close friends such as Francis Boddy and, in my last year, Milton Friedman. Garver himself was a highly competent economist who lacked only one resource, self-confidence. He would ask himself, "If I wrote up my ideas on this subject, would my paper be as profound as the work of Alfred Marshall?" It would have been useless to tell him that there would be no journals in any field of knowledge if we all accepted this criterion of publishability.

In the spring of 1946 I received the offer of a professorship from the University of Chicago, and of course was delighted at the prospect. The offer was contingent upon approval by the central administration after a personal interview. I went to Chicago, met with the President, Ernest Colwell, because Chancellor Robert Hutchins was ill that day, and I was vetoed! I was too empirical, Colwell said, and no doubt that day I was. So the professorship was offered to Milton Friedman, and President Colwell and I had launched the new Chicago School. We both deserve credit for that appointment, although for a long time I was not inclined to share it with Colwell. I nevertheless left Minnesota and went to Brown University, then under the splendid leadership of Henry Wriston. I spent only one happy year there with my old friend Merton P. Stoltz, before I went to Columbia University.

In moving to Brown, Margaret either allowed me or persuaded me to adopt a policy we would follow in later moves: I would buy the new home before she saw it. This procedure had at least one ambiguous advantage for her: I always ended up paying more for the house than I felt that I should, in order to forestall too crestfallen an expression when she first saw her new home. I'm sure I didn't appreciate how tolerant she was. Brown University's original charter gave exemption from property taxes to its professors. By the time I arrived, the exemption had voluntarily been reduced to $10,000 of assessed value and since then, I understand, even this exemption has been abolished. Professors have a few tax perquisites but I regret the loss of this one, especially when the main objection to the tax exemption was that it made us too attractive to rich widows.

The Big City Universities

I have long believed that it is a pronounced advantage for a university to be located in a great city. Such a city is populated with a wide variety of able people: lawyers, merchants, artists, musicians, among the many occupations, and a rich variety of nationalities and languages.

An isolated college tends to become homogeneous—in types of people, types of cultural tastes, in types of activity—in part because university faculties tend to recruit people like themselves. This homogeneity is much weakened when the university is set in the rich sea of talents and temperaments of the city, and the excesses of scholasticism are contained. Indeed there is a danger of the university becoming too enmeshed in the life of the city, as has happened to the universities in Washington, D.C.

It is not an accident, therefore, that many and perhaps most great universities of the world have been located in or close to great cities—thus Berlin, Paris, Harvard, and the like in the western world. When I came to Columbia University in 1947 I felt that I was joining one of the truly great universities, as indeed I was. The next decade, however, was not one of progress for Columbia.

Soon after my arrival, Dwight Eisenhower became Columbia's president, and the appointment proved to be an unwise experiment for both him and the university. He found academic life unattractive. Perhaps it was the fact that a university is a self-governing aristocracy—almost the opposite of the hierarchical structure of a military organization. Or, perhaps it was because he did not find academic life interesting; a university can produce amazingly violent tempests in a teapot. If one is not captured by the fascination of its intellectual adventure, it is a pretty poor business for someone to run. In any event, Eisenhower began taking leaves of absence to make studies of the military establishment, and probably to prepare for his next career.

I do not believe, however, that changes in a large university are due primarily to its president and trustees. They can help or hinder change, but the controlling force must be the faculty itself. If the faculty wants high quality strongly enough, it can get it; and if it doesn't care enough, it will settle for mediocrity. The president, in fact, has little *direct* power. George Beadle, a Nobel laureate in genetics, once remarked that the geneticists at Chicago would not pay any attention to candidates he proposed when he was president of the university.

The Columbia I came to, in any event, was a great university with a strong and varied faculty in economics. Arthur F. Burns, whose later fame was as chairman of the Federal Reserve Board in Washington and then as ambassador to West Germany, was the chief supporter of my invitation, and I saw a great deal of him, more often at the National Bureau of Economic Research than at Columbia. Arthur was a man of vast ability and immense force of personality. He had a superhuman self-discipline in scholarship. His patience and persistence were legendary. One of the National Bureau researchers, Maude Peck, told me that she was hired in 1929 for a few months to help complete a book by Burns and Wesley Clair Mitchell entitled *Measuring Business Cycles*. It appeared in 1947, no doubt reluctantly.

Legend has it that in Arthur's final oral Ph.D. examination at Columbia, James Angell asked him to be a hypothetical Secretary of the Treasury and then posed a dreadful financial debacle with currency flights, bank runs, and general fiscal disaster. What would Burns do? Arthur answered that he would hire a bright young redheaded economist from Columbia (that is, Angell) to cope with the problems, an answer that pleased the remainder of the committee more than it displeased Angell.

Burns once drafted for his annual report for the National Bureau an essay on the role of the consumer in economic life. In the draft he felicitated George Katona, a survey expert and psychologist at the University of Michigan, on the discovery that a consumer's behavior is influenced by other things as well as his income. I expostulated, "For heaven's sake, Arthur, you can't attribute that ancient platitude to a modern scholar!" He told me to wait and see. A few weeks later a well-known economist wrote to Burns and requested to be named as the co-discoverer of this new insight. If I had written the passage, I could not have excluded a tone of mockery and not a single soul would have believed me. In Arthur's years as chairman of the Federal Reserve Board, one friend said he conducted himself as if he could best any other economist in Washington with one hand tied behind him—and he could!

A third episode concerns the oral examinations given to Ph.D. candidates. Each of four examiners was given half an hour to test the candidate's knowledge in four fields of economics. The candidates were often extremely nervous; that is the bad part of the oral examination. The good part is that the examiner could quickly shift from a subject on which the candidate was doing poorly in order to explore other areas in which he or she might do better. Of course the examiners often debated with each other in their successive half hours, using the candidate as a sounding board. In one examination, Arthur started by asking the candidate, a young woman, whether she knew everything about business cycles, and of course she said no. Arthur continued, "Ask me questions for half an hour." What a wonderful examination technique, if only she had known more.

Arthur was a magnificent talker—one cannot say conversationalist. He had original views on many people and many subjects. He spoke with a sort of careful abandon, wholly dissimilar to his measured written words. Many a

National Bureau staffer missed dinner because he stopped in to chat for a moment with Arthur.

Of course I had other outstanding colleagues at Columbia. William Vickrey was already a major figure in public finance and was later to do powerful work on the theory of auctions. Ragnar Nurkse in international trade, Carl Shoup in public finance, James Bonbright in public utilities, Frederick Mills in economic statistics, Carter Goodrich in economic history, and Joseph Dorfman in the history of economic thought were among the members of a strong, varied department. I taught courses in economic theory, the history of economic thought in Europe (for some mysterious reason Dorfman found the thin American literature before 1890 interesting), and industrial organization. I continued there until 1958, when I returned to Chicago (see chapter 10).

The atmosphere of every great university is distinctive. Chicago, the university I know the best, has by some miracle succeeded in remaining young and enthusiastic although it has begun to plan its first centenary in 1992. There is an almost romantic commitment to the pursuit of knowledge that somehow escapes naiveté, probably for the excellent reason that significant bits of new knowledge often appear. A friend who was a professor in our law school, and whose work had been widely acclaimed, told us that he was going to leave the university because he had lost the desire to do research. I joined those who expostulated with him: Couldn't he teach his classes and rest on his laurels? "No," he said, "Chicago isn't a good university in which to retire from scholarship." In a literal sense he was wrong—some people have retired quite comfortably—but in a deeper sense he was right. The inactive scholar may be liked and provide useful and appreciated services, but he or she will not be looked upon as one of the warriors.

The other side of the coin is a warm camaraderie, a willingness by colleagues to share one's problems even if they are not of close relevance to their own work. Drafts of papers are read carefully and constructively, and one is expected to return the compliment (and the sharp criticisms). I understood when a famous economist once said to me, "I never knew a professor could be lonesome until I left Chicago."

The vitality required for this commitment to research is rare. It is all the more remarkable when it is found at a university in numerous groups of scholars over long periods of time. Of course after a time it becomes self-reinforcing; scholars who are attracted to that kind and intensity of scholarship are the ones who accept the invitations to join the faculty. When I returned to Chicago from Columbia I reckoned it a gain that I was coming to an institution where seniority would not be considered an adequate substitute for interesting research.

Such beehives are unusual and therefore probably anomalous. The London School of Economics had this charged atmosphere in the 1930s, and the department of economics at MIT has been equally committed to research during Samuelson's regime. Most universities will be more calm, and more diverse in their goals. Important influence on public opinion is esteemed at Harvard, and in fact it is probably esteemed at most schools but rather less fully achieved. (There is an old Harvard joke that it was hard to get a quorum of the faculty when the Democrats had the White House, and as hard to get a quorum of the Board of Overseers when the Republicans had it.)

One further comment on my university life. Economists have a doctrine in which I place great faith, that of *revealed preference*. It states that people display their true preferences by what they do, not by what they say. Bernard Mandeville, a wonderfully superior philosopher and physician of the early eighteenth century, aptly illustrated the doctrine:

> I don't call things Pleasures which Men say are best, but such as they seem to be most pleased with . . . John never cuts any Pudding but just enough that you can't say he took none; this little Bit, after much chomping and chewing you see goes down with him like chopp'd Hay; after that he falls upon the Beef with a voracious Appetite, and crams himself up to his Throat. Is it not provoking to hear John cry every Day that Pudding is all his Delight, and he don't value the Beef of a Farthing?[1]

When I apply the doctrine to myself, I am sometimes embarrassed by its message. For instance, consider committee life. It is ancient lore in academic life, and probably elsewhere, that committee life is extraordinarily time-consuming and inefficient, because the decisions are reached so slowly and involve so much compromise as to become formless. The attitude is illustrated by the anecdote of Charles Kettering (of General Motors fame) who,

[1] Bernard Mandeville, *The Fable of the Bees*, vol. 1 (1714; reprinted 1924, Oxford, Eng.: Clarendon Press), 151–52.

when told by a lady that Lindbergh's then recent solo flight over the Atlantic Ocean was a marvelous performance, replied, "Madam, it would have been much more wonderful if a committee had done it."

I have, of course, often expressed this conventional attitude toward committee life but "revealed preference" indicates that I began to lead an active committee life at Columbia. I became, among numerous other instances, chairman of the Committee on Instruction of the Faculty of Political Science, and later, at Chicago, spokesman of the Committee of the Council. These were not positions of any administrative responsibility, but rather channels for the faculty to communicate with the administration.

Good communication is essential in the great university. The faculty is too strong to be ignored, and too diversely opinioned to be a conceivable administrative body. Only rational discussion can achieve some measure of agreement. I believe that the excellent machinery of communication was one large factor in the much better history of events at Chicago during the widespread student rebellions of the late sixties than at most other universities, for our faculty united behind the expulsion of a large number of young barbarians. The more radical of the rebelling students of the late 1960s *were* barbarians: They were prepared to suppress free speech and other traditional liberal values with violence in order to advance their intransigent demands. Yet they were less guilty, I think, than the large number of faculty members who, while criticizing coercive tactics, openly sympathized with the students' desire to politicize the universities. Indeed, without this widespread sympathy the student movements would have been much less effective and would not have left a permanent legacy of political engagement and strong partisanship in academic institutions. I cannot omit giving vast credit to Edward H. Levi, first as provost then as president, who is the possessor of astonishing intelligence and superhuman patience, but the machinery of faculty participation was helpful. That machinery had been introduced earlier to deal with a painful conflict between Robert Maynard Hutchins and a large part of the faculty, so I am driven to believe that Hutchins's troubles served a hidden purpose.

These have been my major academic associations.

CHAPTER 3

Economics in Depression and in War

THE SENSE OF DESOLATION of the American economy, indeed of much of the world's economy, in the 1930s is difficult to recapture. Banks were failing by the thousands. I can remember a faculty meeting at the University of Washington in 1933 that lost its quorum within minutes after the rumor of bank closings reached the hall. Unemployment reached historic levels; at its peak one-fourth of the labor force was unemployed. Prices fell to appalling levels: On a trip through California with my parents in that same year, I bought seedless grapes for one cent a pound.

The Great Depression earned its name not only for the depths to which economic activity fell but also for its persistence. America's total real national income did not regain its 1929 level until 1940, when it had to be divided among a population which had increased by ten million. Unemployment did not fall back to the rates of the 1920s until well into World War II. The performance of the economy was deplorable, and the air was filled with denunciations of private enterprise and capitalism.

The study of the causes and cures of depression and unemployment became the central focus of the work of young economists, particularly after John M. Keynes's *General Theory of Employment, Interest, and Money* appeared in 1936. It says a good deal about my own immunity to some of the important intellectual movements in economics that I never became seriously engaged in the study of the movements of aggregate income and employment—now called macroeconomics. This abstention from the subject was certainly not due to my training. Well before Keynes's book appeared, the Chicago economists were urging the need for monetary expansion and public works programs (see J. Ronnie Davis's 1971 book *The New Economics and the Old Economists*). Perhaps I had a subconscious desire to avoid problems of overwhelming complexity; even today there is nothing like an established consensus on the causes of and cures for macroeconomic fluctuations.

The Great Depression made the conditions of economic life the central problem of society in the 1930s, and it is not surprising that it carried the study of economics into an era of prosperity. The prosperity included an intellectual dimension; large numbers of the best graduate students entered economics because it was the discipline grappling with what seemed to be society's most urgent problem. The prosperity was also commercial—at least in comparison with most academic subjects at the time.

Washington, D.C. began to employ economists on a substantial scale, and soon it was teeming with young economists. That demand continued to grow when the nation
eventually emerged from the depression and entered into
war. I doubt whether the number of economists in government declined appreciably even after the ending of
World War II.

Economists were not unique in prospering from the
misfortunes of others: Glaziers are proverbially enriched
by hailstones. An economist must add that the glaziers do
not really benefit from hailstones; taking one year with
another they will earn only enough to keep their numbers up. If they earned more, others would pour into the
occupation and lower earnings; the newcomers would
be *their* hailstorm. In similar fashion, the economists
multiplied rapidly in number in response to their new
prosperity.

I joined the throng in Washington in 1935 for part of a
year, in a suitably humble role as an assistant economist
in the National Resources Planning Board. Whatever the
Board's great, never-to-be-fulfilled tasks were, my own
task was narrow. I was to help estimate the portion of the
benefits of federal works that might properly be charged
against state and local governments because of the local
benefits they received. If a dam or a flood-control project
increased the value of neighboring lands, why, the Planning Board asked, should not these lands be assessed for
these benefits to help pay for the projects? I do not wish
to remember precisely how well or poorly these estimates
were made because I was a rank amateur at this kind of
work. Scraps of our output finally drifted into one of the
Board's reports, but the entire project never reached seri-

ous implementation. There were not many local governments in America in 1935 that could have paid for any appreciable part of the federal public works in their vicinity.

All this attention had its costs as well as its benefits for economics. The intervention of a science into political policy has the effect of forcing it to deal with real problems, and that is to the good. But often a science is confronted with problems it is unequipped to solve. The very source of public interest in economics—the deplorable state of the American economy—implied a dissatisfaction with the received economic knowledge: How could we be in such a predicament if economics had good answers to our problems? This question is not really very cogent; it ignores the fact that the "best" economic knowledge at any one time does not go around wearing a halo so it can be readily identified by everyone. When crises are urgent, new theories to deal with them are proposed by some and accepted or rejected by others without the trial by fire that is the heart of the scientific process. That trial is one of close, persistent examination and hard argument and repeated testing of the ability of a new theory to explain real economic events. It is unusual for an important economic theory to go through this trial in less than ten or twenty years—and who will wait that long to launch a program to increase employment or save failing banks?

Consider the famous book launched in 1932 by Adolph Berle, a brilliant Columbia University law professor, and Gardiner Means, a most inventive economist. It was called *The Modern Corporation and Private Property*, and it argued that stockholders had lost control of the corporations they owned. The corporate executives who ran the enterprises to their own taste had negligible stock holdings; a professional bureaucracy now ran big business in America. The authors cited as a striking example of this separation the fact that in 1930 the twenty largest stockholders owned only 4 percent of the stock of the American Telephone and Telegraph Company, and the holdings of the officers of the company were negligible. The separation of ownership and control, and the implied lack of accountability of corporate officers, became and remains to this day a part of the conventional wisdom.

The difference between the lawyer and the economist in the treatment of evidence is worth noticing. Adolph Berle had asked himself whether corporate management in fact paid itself exorbitant salaries, engaged in dealings that injured stockholders for the private benefit of the management, and otherwise shamefully betrayed its fiducial responsibilities to the stockholders. He was able to cite legal cases, sometimes quite a few cases, in which exactly such practices had been demonstrated in court. Q.E.D.

The economist, on the contrary, believes that the existence of individual cases of scandalous behavior neither proves the importance of illicit self-serving behavior by corporate officers, nor does it justify, say, a new set of political controls over corporations. Twenty-five or even a hundred cases of misbehavior a year are trivial in their impact on events in an economy of several million corporations. If in each of these few abused corporations the officers overpay themselves fivefold, that will not amount to one-tenth of 1 percent of corporate-officer payrolls.

That is only half the story. There should of course be rules to punish unscrupulous corporate officers and keep their number down, and, indeed, the conventional rules on fiducial responsibility and fraud serve precisely these purposes. But no conceivable alteration of corporate law or the Securities and Exchange Commission's regulation will ever eliminate *all* such transgressions. Even if one could devise such draconian measures, they should not be adopted. They would so hamper ordinary corporate decisions as to cost stockholders far more then they saved. In the event, however, it was the lawyers' kind of evidence that carried the day.

The economists' tests were feasible even at the time. The independence of corporate officials should have led them to pay themselves unusually large salaries, but that did not happen. The pay of executives in corporations with widely diffused stock ownership was no higher than in corporations of the same size with concentrated stock ownership. Yet according to Berle and Means there should have been a strong relationship because a highly diffused stock ownership would have made it difficult for stockholders to unite to discipline or discharge the officers. Nor were the corporations with diffused stockholdings less profitable (in 1932 perhaps one would say more unprofitable) judged by the rate of return on assets. These tests did not come until many years later, although much evidence was already at hand in the thirties if anyone had wished to use it.

Dr. Means soon launched a second powerful thesis. The age of market competition was ending, he said, and prices were now being set unilaterally by the giant corporations. Prices were no longer set by supply and demand; they were announced by major producers and maintained for long periods without change. In Means's language, they were *administered*. Since the markets for industrial products were no longer subject to the discipline of competition, shouldn't they be controlled by publicly constituted bodies?

Critics soon poked a hole or two in this argument, but judging by the continued popularity of Means's views, the holes did not seem big. For example, Means argued from the fact that many (published) prices were rigid, and from the assumption that this rigidity was a modern development, that rigid prices were a large cause of the Great Depression. Frederick Mills and Donald Humphrey showed, on the contrary, that these prices had been equally rigid as far back as the statistics went, which was 1890, so rigid prices did not seem like a good explanation for why the 1930s were so different from earlier decades. But again, as with the separation of ownership and control, it was decades before most basic tests of Means's argument were made. I shall argue later, however, that the prime mover of public regulation is self-interest, not ignorance.

I do not want to flatter Dr. Means to the extent of making him the only producer of important economic doctrines in industrial economics during the Great Depression. A separate theoretical development was even more important in influencing economic thought for decades: the theory of monopolistic or imperfect competition. In 1933 Edward Chamberlin published his doctoral dissertation, *The Theory of Monopolistic Competition*, and Joan Robinson, the brilliant Cambridge University economist, published *The Economics of Imperfect Competition*. Chamberlin and Robinson had their differences and their vigorous quarrels, but they also had some common ground. The center of that common ground was the conviction that modern industrial economies such as Britain and the United States were no longer competitively organized; some troublesome measure of monopolistic power was possessed by almost every large firm. They differed in how this market power arose. Chamberlin thought it arose out of the fact that every seller and his product were different from those of other sellers and products, even when all were producing, for example, toothpaste. Robinson thought that market power arose out of sheer size of the firm relative to the market, so General Motors had market power because it produced 40 or 50 percent of the cars, not because Chevrolets and Fords were different. Robinson's book was the more elegant and lucid, Chamberlin's the more concerned with real economic life.

These books and the large literature to which they gave rise reinforced the work of Gardiner Means by emphasizing the prevalence of monopoly and oligopoly (few sellers) in the modern western economies. The economists of the next forty years devoted primary attention to theoretical and empirical problems of monopoly and few sellers. (Means had also launched the statistical study of concentration ratios, which report the share of an industry's output coming from a few largest companies.)

When a fundamental change in the focus of interest of a science takes place, it carries everyone along with it. The new focus (Thomas Kuhn has called it the paradigm) defines the problems worth talking about, and those who are critical of the new approach are influenced as much as its supporters. Even in the 1930s I was skeptical of much of this new literature on monopoly and its prevalence.[1] Nevertheless I, equally with its supporters, wrote vastly more in the years to come about monopoly than about competition. An article vigorously denying the prevalence of monopoly helps almost as much as one urging the opposite to keep that subject at the forefront of professional attention.

During the mid-thirties, however, I had been more occupied with writing my thesis than with corporate governance. I had spent a happy, busy two years as an assistant professor at Iowa State College, and finally finished my dissertation and received the Ph.D. from the University of Chicago in 1938. I left Iowa State with one unusual gift. I had argued in class, as young economists will, that economists are neutral in their treatment of good and bad commodities and, for example, drew no distinction between opium and pork chops. For each proposed difference in one I found a counterpart in the other, and when one student said that there are opium addicts, I acknowledged that I was a pork chop addict. On the last day of class each of the class's thirty members presented me with a handsome pork chop, and my addiction was permanently impaired.

My wife and I moved to Minnesota that autumn and began to raise a family. It was a pleasant Sunday morning in December of 1941 when the telephone rang while I was playing with our four-month-old son. My friend

---

[1] Years later when I was a professor at Columbia University, I attended a meeting of the American Economic Association in Washington, D.C., and on the flight back to New York to my surprise I found myself sitting next to Edward Chamberlin. He opened the conversation, "You and Professor Knight are the two most mistaken economists I know on the subject of monopolistic competition." Thank heaven it was a short trip.

Francis Boddy told me what our silent radio could not, that Pearl Harbor had been attacked.

If economists were dictators in war, I wonder whether they would undermine their positions by having very few wars. On a reasonable cost-benefit calculation we should often reverse the slogan, "Millions for defense, not one cent for tribute." Could not the bloodiest war in our history have been avoided by *buying* the freedom of the slaves, just as Great Britain had bought the freedom of the slaves in the West Indies in 1833?

I do not know how far money can substitute for violence. Counterviolence has the undeniable advantage of imposing heavy costs on the prospective attacker, and thus of discouraging nations from entering the ransom-extortion industry. Then, too, the passions of war make it possible to shift the costs of war among people in a way that one cannot easily, or possibly at all, shift the financial costs of taxation for tribute or ransom. One can conscript the military force at wages that are much below the amounts the conscripts would earn in civil life. It is difficult to believe young adults could be *taxed* at rates ranging from one-tenth to more than nine-tenths of their civilian wages. More fundamentally, how far can money substitute for passion? It is ancient wisdom that war unifies a country (Vietnam to the contrary), but does it do so by bringing about agreement on goals or by silencing dissent?

My first taste of the change in life that most young men were experiencing came in 1942 when I joined a predecessor organization to the Office of Price Administration, already under the opportunistic leadership of Leon Henderson. Henderson had acquired a certain fame in Washington when he had been one of the few to predict the crash of 1937. An indulgent public had forgiven or forgotten his identical but mistaken predictions in previous years. I still label the repetition of a prediction until it comes to pass the "Henderson method."

The Henderson domain contained two divisions. One engaged in jawboning businessmen not to raise prices ("there has been an unwarranted rise in the price of X and sellers should retract their price increases"). The second branch, of which I was a member, was called the Defense Finance Unit and was headed by Raymond Goldsmith. It had the task, as we saw it, of devising fiscal policies to contain inflation. We were opposed to price controls and, indeed, even to jawboning, so Herbert Stein, who was just as clever then as now, urged the staff of the price control branch to sell short the commodities whose prices they claimed were rising unwarrantedly. After all, unwarranted prices soon fall, so there are profits to be made by short sales. Of course we were fighting a hopeless battle, and once price control powers were given to the OPA, we were overwhelmed by the price controllers, who soon procured John Kenneth Galbraith to be their administrator of the price system. A fox in the hen house? I retired briefly to academia.

I was impressed by how utterly unprepared most of us—generals and politicians as well as economists—were to think realistically about the problems of a modern war. The Defense Finance Unit talked of alternative military budgets of two or four billion dollars per year, and from the unit only Homer Jones insisted on adding a zero to such numbers, thus bringing them measurably closer to the real magnitudes of wartime. The agenda of military procurement we were shown, prepared well before 1939, was simply ridiculous—it even maintained the cavalry.

Many years later I received a note from Tjalling Koopmans asking whether I had really said that if Manhattan was bombed, the best way to evacuate the population would be to use the price system. I was taken aback by his letter, because I had not even thought of that problem. But I told Tjalling that the first time Manhattan was bombed *any* system of evacuation would be grotesquely confused and inefficient. If the bombings became repetitive, however, I thought the price system could handle the problem well. The first half of my answer was surely correct, and I believe now even more than I did then in the market system's flexibility, adaptability, and resourcefulness in finding new ways to make money. I pray that the answer will never be tested.

As the war progressed, I eventually joined a research unit called Statistical Research Group at Columbia University. It was headed by Allen Wallis and contained economists such as Milton Friedman and statisticians such as Harold Hotelling, Abraham Wald, Jacob Wolfowitz, and later, Leonard Savage. It was a pioneer American branch of the new craft called operations research, which applied statistical and economic theory to combat problems and to wartime procurement. It worked on problems such as the fusing of proximity bombs, the spacing of torpedo salvos fired from a destroyer, and the armament of fighter aircraft. Our group had illustrious successes, such as the invention by Wald of a new method of statistical analysis called sequential analysis. That method of quality inspection saved the economy more per month in the purchase of rocket propellant than the entire wartime cost of our organization. My role in our work was so modest that my claim must be that I did not aid the enemy.

The few lighter moments in this period often came from mathematicians. A memorandum on the correct strategy to fight Japan emerged from Princeton University, and I congratulated its author on single-handedly rediscovering economics up to 1820. That patronizing note quickly elicited a characteristic blast from John Tukey, who later forgave me when we were fellows at the Center for Advanced Study in the Behavioral Sciences at Stanford. A paper by another, much lesser mathematician stimulated me to write on "Should We Bomb Japan Continuously or Discontinuously?" Allen Wallis deemed it so scurrilous that even my own copy was suppressed.

A lesson I learned from this experience is that one can become an expert in a narrow field with astonishing rapidity. One subject I worked on was the vulnerability of aircraft to various kinds of firepower (20 mm. cannon, .50-caliber machine guns, etc.). Within six months after our group had begun work on this subject, we were consulted by other war-research agencies on the details of aircraft vulnerability. One day I would be measuring a secretary to estimate how many square feet of target a seated pilot made, and a short time later I would be gravely discussing that number with another research group. I should perhaps already have learned this lesson from economics. The great English economist, Stanley Jevons (1835–1882), launched the modern work on the measurement of movements of price levels with a monograph he had written in only a few months. He was measuring the impact of Californian and Australian gold discoveries in 1848 on English prices, and found a "serious fall" in the value of gold of 1 percent a year. Jevons became an instant expert as of 1863 because of his creativity, but usually one becomes an expert simply by specializing sufficiently. When I wrote a little monograph on domestic servants in 1947, I was the first (and probably last) professional economist to write on this declining clan since the 1890s! One becomes *the* expert because no one else thinks it is worth sharing the expertise: Few wish to share a grape.

With the ending of the war in the Pacific, economists speedily returned to their academic posts and their academic interests. Warfare throws up many economic problems, but the problems do not succeed in holding the interest of economists. I do not know of a single major economic study of war that was written afterward in peacetime. In this economists are no different from others: Our work was essentially conscripted, and conscription is a poor way to arouse a love for a calling. I soon left Minnesota for Brown University and a year later Columbia, and I never again left academia for any significant amount of time.

CHAPTER 4

The Strategy of Science: The National Bureau

IN DEALING with intelligent people, one must produce what they want and convince them of that fact. So it is with scholars. Their first task is to develop ideas that advance their discipline. In economics, the ultimate goal is to increase the understanding of economic life: What happens and why. The more immediate goal is often to develop a theory or conceptual system that helps to identify the central forces in economic life and by this route help achieve the ultimate goal. The second task is to persuade fellow scholars that one is right. A brilliant piece of work that no one reads, or, if read, everyone disbelieves, is an abject failure. A theory or idea that is fully accepted by one's fellows is a magnificent success. Indeed, general acceptance is the fundamental test of excellence.

Scholars, though they often seek exemption, are subject to the rule that a man may not judge his own case.

The scientists in a field of study address each new piece of research with a deeply skeptical attitude. The skepticism begins with the question of whether or not one should even publish the article presenting the research (and an even earlier question as to whether the research should be supported financially). The *Journal of Political Economy*, the journal of the University of Chicago's Department of Economics, is one of the profession's leading journals. It publishes perhaps one out of every ten or twelve articles submitted to it, and that one usually only after a significant amount of revision.

Once published, should the article be read? The first ten English-language economic journals in terms of quality publish perhaps 500 to 750 articles a year, and even the most diligent scholar is going to read only a modest fraction of them. (And this count ignores at least another 150 or 200 English-language journals in economics!) To read an article carefully and thoughtfully is often a task of hours, and any article that is read with this thoroughness by even fifty economists is automatically a truly major article. Skepticism is maintained in the reading process. The correctness of the argument is more than a matter of logical consistency—has the right formulation of the idea been achieved? An immense amount of the controversy created by Keynes's *General Theory* was over what he meant, in spite of the fact that Keynes was a distinguished literary stylist.

There remains the still more basic question: Is the idea important? A theory may be logically impeccable, or be capable of being made so, and yet yield no important contribution to the explanation of economic phenomena. An immensely famous theory of profits (the return on capital investment) by David Ricardo can serve as an example. He argued, and the argument can be made rigorous, that the primary factor determining the rate of return on capital in an economy was the cost of raising the food required by the workers. The wages of workers were governed by the workers' subsistence requirements—which then consisted largely of the cost of food—and when wage costs rose then less was left as return to capital. This theory dominated English economics for decades through the influence of Ricardo's follower, John Stuart Mill, but changes in the cost of food were not within rifle range of being the major determinant of wages in the nineteenth century: The dominant determinant was the rising productivity of workers. Notice that here the economists were not skeptical enough, or skeptical soon enough.

So even professors must be persuaded to study new ideas, ideas that are often potentially disruptive or destructive of the established knowledge of these very professors. The opposition can be powerful: Max Planck said that a science progresses by having the old professors die off. I assume that he would have agreed, however, that shooting old professors would not hasten scientific progress.

The individual economist who seeks to change the views of the profession has essentially zero prospects of success unless he or she is gifted with ideas and heavily endowed with energy, as I have already remarked. The ideas will be easier to sell the more congenial they are to the profession, so an extension of familiar doctrine will be much more readily accepted than a basic reorientation of the science. The new ideas will normally require much repetition, elaboration, and, desirably, controversy, for controversy is an attention getter and sometimes a thought getter. I am reminded of a tale about William Wrigley, the chewing gum king. Someone asked him why he continued to advertise his extremely well-known products so lavishly. He pointed to a passing train and said, "Why don't they disconnect the locomotive, now that the train is moving?" It takes a locomotive of sorts to keep an idea moving in a science.

Students are strong allies in this process but it demands a certain measure of success before a professor gets access to good students. There may be an even stronger set of allies: colleagues who share a general viewpoint and agenda of research. When these allies are strong enough or numerous enough, they constitute a "school." I distinguish a school from a master-and-disciple relationship, at least as a matter of degree, by the presence of several major scholars in a school, scholars who may even disagree among themselves upon some matters of importance. But the distinction between schools and master-disciple teams is not always easy to draw. The scholars comprising a school must have a program, a shared view of the proper methods and problems of the science that differs in important (or so it must seem) respects from the ruling views of the scientific community in which the school belongs. A strong and persistent (but not necessarily unchanging) central intellectual core gives cohesion to the members and their work. Yet this intellectual core must not be so narrow or rigid that it cannot comfortably

accommodate the necessarily varied interests and apti-
tudes of a number of able people.

I have spent long periods in two important schools of
economics: the National Bureau of Economic Research
and the Chicago School. I shall discuss the National Bureau here and deal with the Chicago School later. I began
working at the National Bureau in about 1942, and I did
not finally leave it until some thirty-five years later. My
most intimate association with it was during my Columbia days, 1947 to 1958.

The National Bureau was founded by Wesley Clair
Mitchell and others in 1920 as a private, nonprofit institution to establish the main facts about the structure and
operation of the American economy. Mitchell was a superb empirical economist, although unfortunately he had
an inadequate appreciation of the uses of economic
theory. The Bureau pioneered in works of great importance, and the most important was to create the national
income accounts of the United States.

There are two basic ways in which one can measure the
total annual performance of an economy:

1. One can list and add up (by value) all the final products
   turned out in a year: potatoes, chefs' services, new
   houses and automobiles, machinery, doctors' services,
   and so on. The total is national product.
2. One can list and value all of the productive resources
   used to make these products: the infinitely varied kinds
   of labor and the services of capital and natural resources.
   The total in dollars is income earned.

Note that they should sum to the same amount.

First Willford King and then Simon Kuznets overcame thousands of analytical and data estimation problems in order to construct the present system of national accounts, which Kuznets established on an ongoing basis for the federal government in 1933. The full details of the national income accounts are stupendously varied: the earnings of professionals, the tips of waiters, the earnings of gamblers, the fringe benefits of workers, the discounts received by employees of retailers—these are just a few of the components of wage income.

The second main area of the National Bureau's work was on the business cycle, the subject that Mitchell had pioneered. Great strides were taken in describing business cycles and developing treasures of economic data. No reliable explanation was produced for the cyclical fluctuations, however—a failure the Bureau shared with the rest of the world.

A fundamental purpose of the earlier studies was to correct grossly mistaken interpretations of economic life. The national income data made it impossible for anyone but the most vulgarly ignorant to say that "Wall Street" or "capital" gets most of the income of the nation, or, indeed, that the share of income going to wage earners was declining. At least three-quarters of national income went to labor, and the fraction was not declining over time. In the 1930s the Bureau's studies squelched the belief that over-saving by our nation as its wealth grew was a cause of persistent unemployment or depression, for the rate of saving out of income was roughly stable from the 1870s to the 1930s. No verdict has been given on the current complaint of under-saving.

The Bureau's procedure in reviewing research was designed to create public confidence in the results. The care in conducting the research work was the basic source of confidence, and it was instilled by Mitchell, Burns, Solomon Fabricant, and Geoffrey Moore. A board of directors that included academicians, businessmen, and labor leaders reviewed the manuscripts before publication. This procedure did achieve a high level of confidence in the results, and to this day it is the Bureau that retrospectively selects the official dates when recessions and recoveries begin in America. One main weakness of the procedure was that an obstinate and ignorant director could hold up a research publication for years. This happened when C. Reinhold Noyes long delayed the publication of Milton Friedman's splendid analysis of the incomes of independent professionals (in law, medicine, dentistry, and engineering). It eventually appeared, with work by Simon Kuznets, as *Income from Independent Professional Practice* in 1946. Noyes's opposition was aroused by Friedman's argument that the American Medical Association acted as a monopolistic cartel. A second weakness of the Bureau's review procedures was that they almost guaranteed dull books. James Bonbright once remarked that if the Bureau had published the famous Kinsey report on American sexual practices, it wouldn't have sold five copies to the public.

The Bureau was normally strapped for money and Burns in particular conducted its affairs with severe frugality. He finally accumulated a small endowment, and complained bitterly when later directors of research were more relaxed in their spending. He was not consoled when I told him that it is customary for frugal fathers to despair over the conduct of their sons.

My own work while I was at Columbia University was largely in the industries that provide services such as education, government employees, scientific personnel, and trade. It was not exciting. When I did develop an interesting method of determining which sizes of companies in an industry are efficient or inefficient (the survivor method), the National Bureau found it sufficiently dubious to encourage publication elsewhere. The survivor method rested on simple Darwinian logic: Watch an industry over time to see which sizes of companies do well and which do badly. The successful sizes are on average doing well in all the things that matter; inventing new methods and products, dealing with unions and governments, and using efficient merchandising techniques. This method is often used by economists in studies of the efficiency of various sizes of businesses in an industry.

I acquired familiarity with the sources, construction, and use of economic data, and even more I acquired respect for the power of empirical data in testing economic theories. At leading centers of economic theory, such as MIT, it has been the practice to ask: Is the new theory logically correct? That is a good question but not as good as a second question: Does the new theory help us to understand observable economic life? No one will deny the desirability of eventually asking the second question, but many economists prefer to leave that question for a later time and a different person to answer. That division of labor is quite proper, but until the second question is answered, a theory has no standing and therefore should not be used as a guide to public policy.

One lesson that was impressed upon me at the National Bureau was how much more demanding people are for evidence to support new, and especially controversial, ideas. Of course, on reflection that is a perfectly reasonable demand. Familiar ideas have been around longer and their continued currency suggests that they are not easily refuted.

The National Bureau was a school with respect to methodology rather than with respect to desirable public policy. Wesley Mitchell, for example, was at least mildly favorable to central economic planning of a nation (hence his membership on the National Resources Planning Board), while Milton Friedman and I, among others, were strongly opposed to such a policy. The emphasis upon the patient, thorough establishment of facts of evident relevance to scientific research and public policy served a useful purpose. The modern Bureau no longer has this mission, and that is inevitable. As I shall argue, a school must eventually change its mission whether it wins or loses. The modern Bureau has accepted the nature and organization of university research and seeks primarily to assist in supporting cooperative economic research by academic economists.

CHAPTER 5

Eureka!

SCIENTIFIC DISCOVERIES are usually the product of dozens upon dozens of tentative explorations, with almost as many blind alleys followed too long. The rare idea that grows into a hypothesis, even more rarely overcomes the difficulties and contradictions it soon encounters. An Archimedes who suddenly has a marvelous idea and shouts "Eureka!" is the hero of the rarest of events. I have spent all of my professional life in the company of first-class scholars but only once have I encountered something like the sudden Archimedian revelation—as an observer.

First, the setting. The most famous economist in the English-speaking world from 1885 until 1924 was Alfred Marshall. As the Professor of Political Economy at Cambridge University, he held the premier professorship—indeed he made it the premier professorship—in economics. This immensely creative (and crotchety) man invented, among many other things, the category of external economies: things that affect a business but are not appreciably influenced by any *one* business. Suppose, for example, that there are thirty coal mines in an area, and each must pump water out of its lower shafts if it is to be operated. If the land is at all porous, the more water the other mines pump, the less water any one mine need pump, but no *one* mine's pumping has an appreciable effect on the water level. Then the more mines that operate, and the more water each pumps out, the lower the cost of pumping water from any one mine.

Marshall was succeeded in the Cambridge professorship by Arthur Cecil Pigou. Pigou paid vast attention to external economies and diseconomies in his treatise, eventually entitled *The Economics of Welfare*. Since no one mine's pumping has a significant influence upon the level of water in the area, in deciding how much to pump each mine owner ignores the beneficial influence his pumping has on other mines, because he receives no compensation for this benefit. Hence too little pumping occurs from the viewpoint of the industry as a whole. Pigou proposed that such external economies be subsidized by the state. This might be done by paying each firm 10 percent of its pumping costs (if that is what it takes, in technical language, to equate the marginal social cost of pumping to the marginal social return).

The converse case of external diseconomies is often illustrated by a factory with a smoking chimney, the smoke from which raises laundry bills $10 a year for each of a thousand households in the neighborhood. Here Pigou proposed to tax the factory, or more precisely, its smoke. The ideal solution might be to reduce the smoke by half with a benefit of $5,000 a year to the households, at an annual cost of perhaps only $3,000 for smoke reduction.

The disharmonies between private and social interests produced by external economies and diseconomies became gospel to the economics profession. Economists accepted this gospel the way they accept supply and demand as the forces determining prices—instinctively and without misgivings. When, in 1960, Ronald Coase criticized Pigou's theory rather casually, in the course of a masterly analysis of the regulatory philosophy underlying the Federal Communications Commission's work, Chicago economists could not understand how so fine an economist as Coase could make so obvious a mistake. Since he persisted, we invited Coase (he was then at the University of Virginia) to come and give a talk on it. Some twenty economists from the University of Chicago and Ronald Coase assembled one evening at the home of Aaron Director. Ronald asked us to assume, for a time, a world without transaction costs. That seemed reasonable because economic theorists, like all theorists, are accustomed (nay, compelled) to deal with simplified and therefore unrealistic "models" and problems. Still, zero transaction costs are a bold theoretical construct. It implies, for example, that in buying an automobile one knows the prices all dealers charge (with no cost to anyone in time or money), that one is completely certain what all warranties for replacement of defective parts or provision of services mean and has complete confidence that they will be fulfilled (without controversy), and so on. Zero transaction costs mean that the economic world has no friction or ambiguity.

Coase then asked us to infer that in this abstract world there would be no external economies or diseconomies, and I guess rather than remember that we accepted this implication without much argument. The thirty coal mine operators would arrange a contract to have each do the amount of pumping (for suitable compensation) that made industry profits the largest possible, since meetings to arrange a contract and the enforcement of the contract would be free. Indeed such meetings would still be cost-less if there were 3,000 mine owners. In a regime of zero transaction costs, lawyers would perish.

But Ronald asked us also to believe a second proposition about this world without transaction costs: Whatever the assignment of legal liability for damages, or whatever the assignment of legal rights of ownership, the assignments would have no effect upon the way economic resources would be used! We strongly objected to this heresy. Milton Friedman did most of the talking, as usual. He also did much of the thinking, as usual. In the course of two hours of argument the vote went from twenty against and one for Coase to twenty-one for Coase. What an exhilarating event! I lamented afterward that we had not had the clairvoyance to tape it.

The argument turned on a picturesque example Coase had used. A cattle rancher lives next to a grain farmer, and occasionally the cattle of the rancher invade the fields and damage the grain of the farmer. Does it make any difference in the number of cattle maintained and the amount of grain grown, whether the cattle rancher is responsible for the damage to the grain or the grain farmer is responsible? The Coase answer is: No! One way of making Coase's answer plausible is to ask what will happen if both the grain farm and the cattle ranch are owned by the same person. That single owner should combine the two operations to achieve the largest profit. If, for example, adding another head of cattle raises cattle profits by $100 but lowers grain profits by $120, he won't add that head of cattle. Similarly, he will decide on building a fence only if the savings over the years fully compensate for the cost of the fence. But separate owners of the grain farm and the cattle ranch can achieve exactly this best solution by contract, and they will be led to do so because then they will have a larger pie to divide. The assignment of legal liability for the grain damage will determine who pays whom, but it will not affect the best way to conduct grain farming or cattle ranching.

This proposition, that when there are no transaction costs the assignments of legal rights have no effect upon the allocation of resources among economic enterprises, will, I hope, be reasonable and possibly even obvious *once* it is explained. Nevertheless, there were a fair number of "refutations" published in professional economic journals. I christened the proposition the "Coase Theorem" and that is how it is known today. Scientific theories are hardly ever named after their first discoverers (more on this later), so this is a rare example of correct attribution of a priority.

Of course transaction costs are *never* zero. Even a simple trade, such as dollars for German marks, has a cost of 1 to 2 percent of the trade. If you change one dollar into marks, and then marks back again into dollars, you'll be lucky to end up with ninety-six cents. If you exchange $1,000 for marks, and then marks back into dollars, you will end up with more than $960 because lesser charges are made on larger sums. There is a cost to running a currency exchange and the tourist pays for it. The Coase Theorem is nevertheless of great interest because the magnitude of the transaction costs puts a ceiling on how large external economies or diseconomies can be.

If one is not an economist, one might well be inclined to say that Coase had a cute point, but why the excitement? A world free of transaction costs is, as you say, a farfetched state of affairs, and it should be expected that it leads to bizarre results. Think of all the odd things one could imagine of a society that did not obey the law of conservation of matter, and shrank with every bite!

I must say in candor that there is some point to the remark. Theorists like new and strange constructs that create a new world or change the way of looking at the current one. Of course we now have begun to study the nature and size of transaction costs—something we did not do before—but I confess that surprisingly little of this work has been done in the nearly three decades that have passed since the Coase Theorem was published.[1] The reason for the excitement was somewhat different: This formulation immediately changes the way one looks at and studies a vast variety of economic problems. Here is one example.

The "reserve" clause in professional baseball for a long time gave the right to control a player's services so long as he remained in baseball to the team with which a player first signed. (The clause has been eliminated in recent times.) Hence if a team succeeded in first signing a future star, a new Ted Williams for example, it used to be said

---

[1] Ronald Coase, "The Problem of Social Cost," *Journal of Law and Economics* 3 (October 1960): 1–44.

that the team could keep him permanently—even though he might be much more valuable (in terms of gate receipts) to another team. The Coase Theorem tells us that is wrong: The player would be sold to the team where he was most valuable. *That* isn't much of a trick in contract negotiation; both teams and the player would gain by the transfer. The reserve clause could affect how much of his contribution to the gate receipts this player received, but it would not lead to his use on the wrong team.

Examples of the Coase Theorem could be multiplied, but the baseball player's contract should suggest the source of the excitement. The way one looks at many problems is now changed; we look to see whether solutions that benefit both parties can be achieved by negotiation. That approach to problems has wide-ranging effects. Consider the suits at law to settle controversies: Why shouldn't the two parties to a contract dispute agree on a compromise, and save the money both would have spent on a lawsuit? Generally, indeed, they will, and only a tiny fraction of legal disputes reaches the courtroom. Most courtroom litigation is now attributed to large differences between the parties in their estimates of the probability of winning the suit. Where the facts and the law are clear, the suits will be settled before trial. It is not surprising, therefore, that the Coase Theorem has filtered into the leading law schools and receives much attention in courses and law-journal articles on torts, property, and contracts. Coase himself became a professor of economics in the Law School at the University of Chicago, a proper home for his genius.

Creativity often, and perhaps usually, consists of looking at familiar things or ideas in a new way. My most important contribution to economic theory is probably the treatment of information as a valuable commodity that is produced and purchased. I was led to the problem of information by noticing—as every shopper in history has noticed—that one can often find a lower price by canvassing more sellers. Yet the standard theorem in economic theory was that under competition there will be only one price for a homogeneous good in a market. The theorem simply said that both sellers and buyers would seek out and eliminate all differences in prices. For example, as long as a single seller will accept a lower price than the one the buyer was about to pay, the buyer will seek him out. In seeking to reconcile with the theorem the observed array of prices for even identical goods (say, a specific tool priced in thirty hardware stores), I sought an obstacle to complete search for better prices, and found it in information costs. There is a cost in time and travel in discovering how much each seller is asking and how well he provides other services such as a good inventory and quick replacement of defective products. The result is complementary to Coase's work because I had been examining a major component of transaction costs and, in fact, my article appeared at about the same time as his.

The role of information in economic theory has expanded rapidly. There have been single years in which a dozen or more articles appeared on the problems that arise when one party (say, the seller) has more information about the quality of a product than the other party (here the buyer). George Akerlof presented an interesting theory of "lemons." Suppose that only sellers of used automobiles know their quality. Potential buyers will correctly assume that the owners of inferior-quality automobiles will be especially eager to sell their cars, so these potential buyers will offer only as much as the poorest quality car is worth—and that will be the only quality that sellers offer for sale. (This vicious circle can be broken if sellers can acquire deserved reputations for reliability.)

Phillip Nelson, a former student of mine, has made basic extensions of the theory of information to advertising. (Although Phillip wrote his doctor's thesis with me, he never took a course of mine—probably because he had information.) For example, he has shown that so-called "institutional advertising" ("our firm is the oldest or largest in the industry") conveys to the consumer an assurance of reliability: The firm must have treated customers pretty well or it would not have had so many repeat buyers. The economics profession has become a good deal more discerning in its discussions of advertising than it was a few decades back.

Because scientific discoveries often arise when a familiar phenomenon is looked at differently, it is not surprising that many scientific theories have been discovered independently by several people. Often the leading edge of research in a field runs into an obstacle, and several scholars in the field may solve it. Indeed, even these multiple discoverers may actually be rediscoverers. For example, in 1871 Stanley Jevons and Carl Menger (and three years later, Léon Walras) proposed that consumer behavior could be understood as the endeavor of each person to obtain as much satisfaction as possible from the goods he purchases with his resources (the marginal utility theory). This introduced into economics what is called "the marginal revolution." A sadly neglected minor German civil servant, Heinrich Gossen, had presented the theory in a better version some seventeen years earlier, and there were other, earlier anticipators.

It is not surprising that priorities become disputed under these conditions. Robert K. Merton, the great figure in the sociology of science, once gave a fascinating lecture on the importance of priority to a scholar, how the struggle for acknowledged ownership of an idea had turned otherwise calm and generous scholars into ruthless combatants. He illustrated his thesis with many historic examples, such as the struggle between Newton and Leibnitz for the credit for the first discovery of the differential calculus.

In the modern world these struggles no longer are violent or protracted, and, indeed, in economics there has been no really interesting major priority quarrel in the present century nor possibly in the nineteenth century. There have been frequent claims of being a neglected discoverer (thus both Albert Hahn and Michal Kalecki are hailed as anticipators of important elements of Keynes's *General Theory*), but there has been no out-and-out confrontation, let alone charges of major plagiarism. It is my impression that the same toning down of priority issues has occurred in other sciences. One important reason for this change is that the professional journals are now *the* important channel of communication among scholars, so the date of ideas has become easier to determine.

It is almost inconceivable that any modern idea or theory has not had even partial anticipations. History is full of intelligent, creative, curious minds that have grappled with most real problems and many imaginary ones. A modern scholar will be one of a number of able people who is engaged in the cooperative study of a particular area or subject matter, and small pieces of new knowledge (both theoretical and empirical) are frequently added to the common pool of knowledge. The inherited pool of knowledge will be the indispensable foundation on which the next discoveries will be made. This does not imply that the new discoveries are inevitable; every science has had periods of either stagnation or slow and unexciting progress.

The interdependence of scholars and their common reliance upon the same inventory of accomplished research have grown stronger with time. In economics, research moved into the universities in England (then the great leader) in the latter half of the nineteenth century, and the chronology is similar in the United States and western Europe. The number of economists multiplied beyond the fearful conjectures of Malthus. The number of American Ph.D.'s in economics conferred in 1900 was perhaps 10; and in 1980 it was 677.

With this increase in numbers, the economics journals began to appear: Harvard University's *Quarterly Journal of Economics* in 1886, the University of Chicago's *Journal of Political Economy* in 1892, and the *Economic Journal* of England's Royal Economic Society in 1890, and we observed earlier that today the number of English-language economic journals is on the order of 150 or 200. Communication of research has increased apace, so it would be difficult for an economist at a major university to remain ignorant of important discoveries, or to be thoroughly informed about many of them.

The growth in the number of scientists and the vast improvement in communication lead one to suspect that fewer discoveries will be the heroic forward-jumping leaps by individual scholars. It is suggestive that Nobel prizes in physics and chemistry are increasingly awarded to two or three recipients, and it is commonly remarked that the practice is not carried far enough. From 1901 to 1920, thirty-one awards in physics and chemistry were to single recipients, three to two recipients, and one to three recipients (the maximum permitted). In the years 1961 to 1980, the numbers were eighteen to one, twelve to two, and eleven to three—an *average* of almost two recipients. One could ask each prize winner: Would your achievement have been possible without the work of others who did not receive a prize? I suspect that the honest answers will soon be mostly "No."

A discussion of the origin of major or even minor ideas in a science inevitably romanticizes the nature of scientific work; it becomes a tale of heroic advances or brave and costly endeavors. Important individual advances and bold gambles are so tiny a fraction of scientific work, however, as to be comparable to the crust on the tip of the iceberg. Scientific research is a market process, differing vastly in form but little in substance, from the comparable activities of grocers or manufacturers of computers. Individual scholars distribute themselves by the action of self-interest. If macroeconomics or computer science begins to boom, new graduate students will begin to enter these branches in rising numbers.

Scholars are not usually paid by the piece for their research, but they are compensated in equivalent ways. The better the research work, the more prestigious the journal in which it will appear. The superior researcher is hired by the better university, promoted at a rapid rate, favored by the National Science Foundation and the private foundations, given a lighter teaching load, and in the physical sciences supplied with a well-appointed laboratory. The learned societies (as the scholarly trade associations are known) elect the superior scholars—and the more skilful academic politicians—to their offices.

Who decides what subjects to work on, and how good each research product is? In the short run—from year to year—the judges are the fellow scientists. If the subject is at all complicated, the legislators, the bureaucrats, and the academic administrators do not understand the ongoing research, let alone know how to steer it in the most promising direction. The governors of science are the self-perpetuating and self-selecting group of leading practitioners.

A system of government of a science by a self-chosen elite has the potentiality for stagnation and scholasticism of the sort that stultified Oxford and Cambridge Universities in the eighteenth century, but in the American context these evils are escaped. There are so many good universities (not under one control), so many foundations, so many scientific journals, that each major discipline becomes a competitive industry. No one model of orthodoxy can be imposed on a science, although the potentiality of centralized direction has increased since the federal government has taken a powerful role in the financing of research.

The late Joseph Ben-David, a distinguished sociologist of science, attributed the rise of several American disciplines to world leadership to the fact that American science is less centralized than that of France or Germany. A competitive science is more open to new ideas than a tightly organized scientific establishment. The establishment opposes utterly original ideas not only for the good reason that they are usually wrong, but also for the bad reason that when the new ideas are right they render obsolete or irrelevant knowledge of the leaders of the establishment.

The process yields substantial consensus. The year (1957–1958) I was at the Center for Advanced Study in the Behavioral Sciences at Stanford University, my fellow economists were Kenneth Arrow, Milton Friedman, Melvin Reder, and Robert Solow, a group whose most common characteristic was high ability. The Director, Ralph Tyler, asked us to classify a long list of possible future fellows to the Center into four groups, ranging from "by all means" to "never." We did so, and our rankings were so similar that he accused us of collaborating on our selections, although of course we had not.

In a longer run, a science is not self-perpetuating in its composition and problems. Sooner or later those in power will ask for results or better results. The results need not be simply utilitarian (cure disease X or depression Y), but they must satisfy a significant element of the society that they are worth their cost. I conjecture that even here competition plays a large role: competition among disciplines to solve important problems (illustrated by the merging of physics and chemistry), and competition of rival national sciences. German and French economics were dominated for several generations by nontheoretical traditions (historical analysis in Germany, sociological analysis in France), and both have been giving way to the more successful English-American economics.

Eureka!

Economists have been remarkably successful in selling their product in this century. Every large business has at least one economist, and the largest business of all has a Council of Economic Advisors, which on suitable occasions is allowed within the White House. The only addition to the list of Nobel prizes since its founding has been economics. Economics is often the most popular undergraduate major in colleges: It is believed to equip one for law school or business school. All of the tedious humor about the differences of opinion among economists (five economists will have six opinions, two from Keynes), or their infatuation with abstract thinking ("it's all right in practice but it won't work in theory"), are really envious jibes. Denunciation of America is almost the only bond that unifies European intellectuals, and criticism of economics is the chief bond joining the other social sciences. How much sweeter is envy than pity.

In 1982 I was awarded the Nobel Prize in Economic Sciences for my work on the economic theory of information, which I have just sketched, and on the theory of public regulation, which will be described in chapter 7. No rumor had reached me, or rather, only rumors of the selection of other economists, so the award came as a marvelous surprise.

The celebration in Stockholm is superb pageantry. Three sons and their wives and four grandsons, my old friend Walter Bean, and Claire Friedland accompanied me to the ceremonies. People must all have an atavistic reverence for royalty, especially when it includes a charming and attractive queen.

I have puzzled over the role of such a prize and the reasons it commands enormous prestige, especially after reading Harriet Zuckerman's interesting 1977 study of the Nobel prizes, *Scientific Elites*. Alfred Nobel hoped that the generous prizes would provide the recipients with sufficient economic independence to allow them to devote their subsequent careers to research. Yet even in 1901 the initial prizes ($42,000) did not provide economic independence, and today's prizes (often divided among two or three recipients and taxed since 1987) are at best only about three or four times the annual salary of a major scholar. Of course other income flows from the grants—in economics, lecture fees and invitations multiply—but this income must be earned.

The prize is not a device for calling important scientific work to the attention of the scientist's fellow workers. By the time the prize is awarded, an average of about thirteen years after the prize-winning work was published, competent members of the science will know and be using the new results. Only a scientific incompetent would find the award newsworthy for research purposes.

The prize is an added incentive to enter the fields in which it is given. If Adam Smith is right in his belief that men overestimate their chances of winning great prizes in lotteries of all sorts, the Nobel Prize may even have served to increase slightly the number of able young scholars entering the covered fields (physics, chemistry, physiology-medicine—now interpreted to cover biology —literature, and economics) and to lower slightly the average earnings in these fields. It would be difficult to argue that this redistribution of talent is socially useful, however; there is no reason to believe that the covered fields were drawing less talent, and the uncovered fields more talent, than served to equalize prospective marginal products.

The main effect of the prize is to endow the recipients with a large measure of prestige among non-scientists. In this respect the award has been a phenomenal success. The annual ceremonies give the Swedish community a publicity that must be the envy of every advertising executive. For the average educated citizen there is no possibility of understanding the work that won the prize or of tracing any connection between that work and contemporary well-being. Even the uneducated citizen knows that the laureate is a Life Baron of science.

But what social service does this prestige serve? That the prestige invites mild harm is evident: A full collection of public statements signed by laureates whose work gave them not even professional acquaintance with the problem addressed by the statement would be a very large and somewhat depressing collection. But what is the gain from the public esteem?

This is a question, not a veiled complaint. The public has good reasons for what it does, and it is the task of the social scientist to discover them, even though many find it irresistibly attractive to instead ridicule the public's behavior. My conjecture is that the public wishes to admire superior performance in every legitimate calling, athletic and military (for example) as well as scientific. If there were a good objective measure of scientific performance —say, the number of subatomic particles discovered— they would use this basis for selecting champions, rather than the more fallible choices of the Swedish Academy and other learned bodies. But presently it is the best ranking they have, and so they heap their kudos on the laureates.

Why does the public wish to acclaim superior performance in various fields? Does it have a fund of admiration it needs to use up? To continue my conjecture, the acclaim is intended to stimulate truly major accomplishments in the various fields. Major scientific achievements are usually the result of high-risk work. The incentive structure to elicit major achievement in academic life has always relied primarily upon prestige and research facilities. Even the best-paid professor in the fifty leading universities seldom receives three times as much salary as the worst-paid professor. An institution that focuses the prestige of superior work on a few people is a helpful corrective to the egalitarian structure of universities and nowadays of society, and an incentive to undertake high-risk research.

From this viewpoint it is not a major defect of the Nobel Prize that the laureates do not keep pace in number with the relevant disciplines. What is an important defect is the exclusion of fields of comparable intellectual severity and historical grandeur. Thus, at least until recently, a giant such as Laplace would not have been eligible for a prize because celestial mechanics was an uncovered field. Do the various other awards (such as the Field Prize in mathematics) sufficiently fill this need?

CHAPTER 6

Monopoly

IF ONE were to canvass all the books and articles in economics up to 1900, a task best left to the imagination, I believe that one would conclude that monopoly was not a serious problem throughout recent centuries. Alfred Marshall's *Principles of Economics*, the most advanced work in general economics in 1890, devoted one chapter out of fifty-five to monopoly, although he had much more to say about it in his later *Industry and Trade* (1919). Indeed, until 1850 in England the word *monopoly* was usually restricted to describing the exclusive rights to a trade which were conferred by Parliament.

Toward the end of the century attention to monopoly grew rapidly. I am reminded of an episode reported by Lincoln Steffens, a famous journalist muckraker, best known for his expose, *The Shame of the Cities* (1903). He devoted a chapter of his autobiography to "I Make a Crime Wave" in New York City. The crime war arose inadvertently out of a rivalry between two newspaper reporters as to which could report more crimes each day. Finally, the head of the police board, then Theodore Roosevelt, persuaded the reporters to end the journalistic battle and New York City resumed a more law-abiding character. Something like that happened in the United States with monopoly: A wave of muckraking and reform literature commonly emphasized the ruthless power and vast profits of Standard Oil, the railroads, the Steel Trust, and a dozen other corporate culprits.

Economists no doubt read this literature, but they were also influenced by several other developments:

- America passed an antitrust law in 1890, and eventually that law produced mountains of material on business organization and practice;
- there were large merger (trust) movements in Great Britain and the United States at the turn of the century;
- the growing socialist critique of capitalism emphasized monopoly; "monopoly capitalism" is almost one word in that literature;

and because of one more thing: the word changed its meaning.

Throughout most of history, if one saw five separate business enterprises in one industry, say, pressing olives, operated by Adam, Brian, Chester, David, and Edward, one would say "competition." (And possibly also, "What, no Greeks?") Were the five not in rivalry? Brian offered discounts to good customers. David made purer olive oil, Adam developed a higher-yielding olive tree, and so on. There might be periods of quiescent coexistence and periods of aggressive rivalry, but the industry would be called competitive.

The members of a science persistently seek to define its concepts more precisely, and this is all the more true when the science becomes the full-time study of a set of academic practitioners. In the last third of the nineteenth century this process accelerated as economics became a nearly respectable academic subject. (Harvard appointed its first professor of economics, Charles Dunbar, in 1871 and conferred the first Ph.D. in economics in America in 1875 on Stuart Wood, later a businessman in Philadelphia.[1]) One of the concepts that underwent increasing refinement was competition.

One needn't be a hairsplitter (though it helps) to worry about whether competition exists in effective measure if there are only a few business firms in an industry. Why couldn't and wouldn't they agree to set highly profitable prices, especially if they did not fear the appearance of new rivals? And suppose, as I believe to be the case, agreement is unlikely to work at all well with ten separate firms, and hence with more than ten, what about independent rivalry if there are only two or three?

So economists said (to each other), let's be certain that there is competition by requiring a vast number of rivals in a market, and call that kind of market *perfectly* competitive. Surely 100 firms are plenty, and sometimes possibly two are enough, but for rigorous economic analysis let us assume that there are thousands of rivals in the market.

---

[1] Joseph Schumpeter said of Dunbar, "And so, though not a great economist in the sense appropriate to this book, he was a great economist in the sight of God" in his illustrious *History of Economic Analysis* (New York: Oxford Univ. Press, 1954), 866. What a range of acquaintances Schumpeter had!

(In addition, they—we—assumed that all economic actors, such as consumers and entrepreneurs, had complete knowledge of prices and qualities of goods.) Definitions do not yield any knowledge about the real world, but they do influence impressions of the world. If only markets with a vast number of traders are perfectly competitive, and if markets with few traders are called oligopolistic (literally, "few sellers"), that suggests that these latter markets are not competitive, as well as not perfectly competitive.

This suspicion of small numbers was gradually reinforced by the antitrust cases. The first such indictment to be upheld by higher courts was the Addyston Pipe Case. Six firms making heavy pipe for city water and sewage systems combined to fix prices. For example, they would assign Columbus, Ohio, to one firm and the other firms would bid higher prices than the chosen winner when Columbus sought to buy soil pipe. The famous opinion holding this behavior illegal was written by Judge (later President) Taft in 1898.

Of course one such case, or a hundred such cases, would not prove that "agreements in restraint of trade" are prevalent or even common when the number of rivals is small, and there were not a hundred such cases for many decades. In fact, the Addyston case arose because a discharged employee of the pipe manufacturers had made off with the minutes, and gave them to the Department of Justice, and that doesn't happen every day. (If it did, then collusion would stop, so still it wouldn't happen every day.)

The attention to monopoly grew between the wars. My one time colleague at Columbia, Arthur R. Burns (who necessarily shared one mailbox with Arthur Frank Burns, the later chairman of the Council of Economic Advisors) wrote an influential book in 1932 with the ominous title, *The Decline of Competition*. The Temporary National Economic Committee (1938–1941) produced 45 volumes of studies and 33,000 pages of testimony on the evils of monopoly, and armed Thurman Arnold to revive the nearly moribund Antitrust Division of the Department of Justice. Monopoly had become as popular a subject in economics as sin has been in religion. There is a characteristic difference: Economists are paid better to attack monopoly than the clergy are to wrestle with sin. It is to be observed that the economists who defend monopoly in antitrust cases are better paid than the government's economists: Do sinners always earn more than the virtuous who combat them? Probably yes; one must be compensated for bearing the opprobrium of sinning.

A later development made competition seem less likely in many cases where the number of rivals was large. We have already encountered Edward H. Chamberlin, author of *The Theory of Monopolistic Competition*, who argued, as his title suggests, that almost all markets are hybrids of competition and monopoly. Ford and Chevrolet cars are not the same, although they may provide the same transportation and even sell for the same price. Some people would continue to buy their favorite make of car even if it became 10 percent more expensive than the other. If someone were intent upon getting a Chevrolet, Chamberlin would argue that the twenty Chevrolet dealers in a large city are still not in full competition; one may have a handier location, or have a nicer smile, or have a preferred skin color, so the consumer will patronize that dealer even if he charges somewhat more. Since precious few markets are characterized by absolute identity of every specimen of the product traded, precious few markets, perhaps none, are perfectly competitive—so said Professor Chamberlin.

If I sound a little skeptical of the importance of a dealer's smile or skin color, that's because I am. Prices and qualities of goods have primary sway over most markets. Some buyers will be faithful to enterprises for reasons other than price, quality, and service, but they are usually too few and too shallow in their faith to allow an enterprise to prosper if it fails to perform well in the basic functions of business.

Chamberlin devoted his entire life to the support, defense, and modest elaboration of his book (which was based upon his doctor's thesis). Legend has it that when he taught economic theory, he made a determined effort to cover the broad field of price theory, but always ended up concentrating upon monopolistic competition. He had created the idea, and in the end it possessed him. Still, his book made a great splash in the 1930s, and permanently increased the attention that economists pay to the differences among similar products (which Chamberlin labeled "product differentiation").

Great Britain has been less preoccupied than America with the monopoly problem, but British economists shared the growing concern in the 1930s. At Cambridge University in 1933, Joan Robinson had launched her career with *The Economics of Imperfect Competition*, which presented a lucid and elegant theory of monopoly price determination. The extensive controversy that emerged among the supporters of monopolistic and imperfect competition over who was "right" helped to imprint the monopoly problem on economics.

Until the 1950s I accepted the prevailing view of my profession that monopoly was widespread, although I never believed (as many economists did and perhaps still do), that monopoly was the predominant form of market organization in the manufacturing and mining industries. I was an aggressive critic of big business, perhaps leaning a little toward the interesting view of antitrust policy that the now celebrated Judge Robert Bork reported in his book *Antitrust Paradox*. One prominent lawyer (and later associate justice of the Supreme Court) said that antitrust was in the tradition of the sheriff of a frontier town: "He did not sift the evidence, distinguish between suspects, and solve crimes, but merely walked the main street and every so often pistol-whipped a few people." He needed to pistol-whip a few cowhands from time to time. It wasn't important whether they were troublemakers at the moment, for the community at large would be reminded of the presence of law and order.

Some of the flavor of my views at the time can be gained from one of my few appearances before a congressional committee. Edward H. Levi, who was counsel to the Subcommittee of the House Judiciary Committee in its study of monopoly power in the steel industry, asked me to testify in 1950. Here are a few samples:

MR. LEVI: If your advice were followed, and the United States Steel Corp. were broken into segments, would that solve completely any monopoly competition problems?

MR. STIGLER: I think it would solve the monopoly problem at the level of the employer. There would still be monopoly among employees. I personally would favor the application of the antitrust laws to the steel workers' union as well as to the employers.

THE CHAIRMAN [Emanuel Celler]: It would not involve too great a difficulty to truncate from United States Steel its Geneva plant out on the coast?

MR. STIGLER: I can see no problems other than political. As a matter of fact, historically, almost every United States Steel plant existed once independently before it existed as part of the Steel Corp.

THE CHAIRMAN: It would not involve too much difficulty to say that the Geneva plant should be separate and be owned by a separate company and that the Gary plant should be owned by a separate company, and that the plant in Birmingham, Ala., should be owned by a separate company?

MR. STIGLER: I see no difficulties.

THE CHAIRMAN: Do you think we should reexamine our antitrust legal fabric to see whether or not that fabric should be amended so that any excesses or evils that result from these oligopolies could be treated?

MR. STIGLER: Certainly, it is my belief—speaking as an economist and not as a lawyer—that the basic trouble with the anti-trust laws (whether it is due to the statutes or the Court or the Antitrust Division) is their failure to recognize that when an industry is made up of, or dominated by, a few big firms, there is inherently a structure inconsistent with efficient competitive operation.[2]

I now marvel at my confidence at that time in discussing the proper way to run a steel company; I certainly would not presume today to have that knowledge about any industry, even higher education. What is still more embarrassing is that I no longer believe the economics I was preaching, as I will soon explain.

Yet the argument, which I advanced so presumptuously, that U.S. Steel could not owe its size to its superior efficiency has received a strong historical endorsement. If one had bought $1,000 worth of its shares on the last day of 1950 and thereafter reinvested all dividends in the company, the holdings on the last day of 1986 would be worth about $10,600. If one had invested that same $1,000 across the board of the New York Stock Exchange and reinvested dividends, it would have grown to about $50,400 in the same period. I wish I had been able to persuade my father to sell his shares in the company.

So, in 1950 I believed that monopoly posed a major problem in public policy in the United States, and that it should be dealt with boldly by breaking up dominant firms and severely punishing businesses that engaged in collusion. The justification for my fear of monopolistic behavior of the steel industry was the traditional one in economics: The industry was "concentrated." U.S. Steel made about 30 percent of sales of the industry, Bethlehem a little less than 15 percent, Republic about 9 per-

---

[2] George J. Stigler, Hearings, Subcommittee on Study of Monopoly Power, House Committee on the Judiciary, April 17, 1950 (Washington, D.C.: U.S. Government Printing Office, 1950).

cent, and Jones and Laughlin about 5 percent—so the top four firms had 60 percent of output, and much higher fractions of products such as structural steel. Economists (including me) generally believed that this level of industry concentration allowed a substantial amount of noncompetitive behavior, but the belief rested more upon consensus than upon evidence. I held to this position for perhaps another decade, and then my fears of monopoly began to recede. The reasons may be worth recounting since most of them also influenced the economics profession as a whole to move some distance in the same direction.

At all times there have been defenders of private enterprise, whether monopolistic or competitive, and in modern times it was probably Joseph A. Schumpeter who first raised some serious doubts about the importance of the monopoly problem. This Austrian professor at Harvard University was so fascinating a man that I must pause to say a few words about him.

Schumpeter had launched a brilliant academic career just before World War I in Vienna. He is said to have made a vow to become the best economist, horseman, and lover in Vienna, and later to have remarked that he had never fully achieved that degree of mastery on a horse. After various vicissitudes, including a brief term as Minister of Finance during Austria's hyperinflation, he came to Harvard in 1932. Schumpeter was immensely learned and immensely clever, and as a minor foible, a poseur in abstract economic theory. I met him first in 1940 at the meetings of the American Economic Association in New Orleans. He greeted me cordially (characteristically he had read the little I had written) and he soon asked me: "Are you not reminded, dear colleague, of general equilibrium theory in economics when you read modern mathematical physics?" I doubt that I had the courage to admit that I had sadly neglected mathematical physics, and I surely did not dare tell him that I doubted his own knowledge of that area.

In *Capitalism, Socialism and Democracy* (1942), Schumpeter painted an unconventional picture of the capitalistic process. The competition between the Pennsylvania and New York Central Railroads, he argued, might be sporadic and even trifling, but the competition to railroads provided by new transportation media such as trucks, automobiles, and airplanes really mattered. In his words,

[The] competition which counts [is] from the new commodity, the new technology, the new source of supply, the new type of organization (the largest-scale unit of control for instance)—competition which commands a decisive cost or quality advantage and which strikes not at the margins of the profits and the outputs of the existing firms but at their foundations and their very lives. This kind of competition is much more effective than the other [price competition of similar firms] as a bombardment is in comparison with forcing a door. . . .[3]

We economists mostly rebelled against such heresy, but it left its mark.

In my case a second individual had a stronger influ-

---

[3] Joseph A. Schumpeter, *Capitalism, Socialism and Democracy*, 3d ed. (New York: Harper & Brothers, 1950), 84.

ence. When I moved to Chicago from Columbia University in 1958, I began to see much of my friend Aaron Director. He taught with Edward Levi (the same Edward Levi who had served as Emanuel Celler's special counsel) a course in antitrust law, and in the process his luminous mind changed the way in which the Chicago School thought about industrial problems. An example: It was established legend that the old Standard Oil Company had achieved its dominant position primarily through predatory tactics, such as bribing the employees of rivals, actually sabotaging rivals, and especially engaging in local price cutting. That last-named technique was employed when Standard Oil cut prices below costs in a particular town until the local producer was bankrupt, and then John D. Rockefeller and his associates would buy up the rival for a pittance.

Director was skeptical of this tale, and among the reasons were: (1) price wars are costly even, or rather especially, to the victor (whose output and losses exceed those of the smaller rival), and (2) new rivals would appear once the price of oil was raised again to a monopolistic level in that particular town. An able young colleague, John S. McGee, undertook the careful review of the record (some 13,500 pages) and sure enough, that record revealed no important evidence of predatory price cutting: "I am convinced that Standard did not systematically, if ever, use local price cutting in retailing, or anywhere else, to reduce competition."[4] Instead the company usually found it profitable to buy out rival concerns at remunerative prices.

[4] John S. McGee, "Predatory Price Cutting: The Standard Oil (N.J.) Case," *Journal of Law and Economics*, 1 (October 1958): 168.

Monopoly

Aaron Director takes economics seriously. In each of the various practices he has analyzed (tie-in sales, patents, resale price maintenance, etc.) he has sought the profit-seeking reason that led businessmen to adopt the practice. Sometimes the reason was the exercise of monopoly power, but other times an important efficiency was achieved by the industrial practice. Monopoly receded from its near-monopoly position in explaining business behavior. The researches of Sam Peltzman, Harold Demsetz, Lester Telser, and others reinforced the decline in the role assigned to monopoly.

I was also influenced by that nefarious tool of the devil, economic theory. Let us revert to the industry consisting of five pressers of olive oil. Let us assume that each is pressing 2,000 loads of olives at a cost of 1.6 drachmas per load and selling them at a monopoly price of 4 drachmas per load, and reaping monopoly profits of 2,000 times 2.4 (4 − 1.6) drachmas or 4,800 drachmas per season. If one of them, say Chester, can secretly cut the price to 3.5 drachmas and capture the business of one of the largest olive oil buyers, worth, say, the oil from 400 loads, he will gain 400 times 1.9 (3.5 − 1.6) or 760 extra drachmas of monopoly profit. How can he be detected? Not so easily: Neither Chester nor his favored customer has a clear incentive to reveal the arrangement. The article I wrote on this problem, "A Theory of Oligopoly," dealt with the detection of cheating and concluded that often cheating could not be reliably detected. That makes conspiracies harder to form and weaker if they are formed. This line of argument was an extension of my work on the economics of information.

Moreover, the facts seemed to confirm this conclusion.

The study that impressed me the most was the one made of the underwriting syndicates that are formed to buy and resell the bond issues of state and local governments. When Texas or Detroit or the Washington Nuclear Power Consortium (it better not try again) wants to sell bonds, it advertises for bids. A group of underwriters combine and bid some price such as 99 for a ten-year bond with a coupon rate of 5 percent (the bonds are tax exempt). Sometimes only one consortium bids, sometimes as many as twenty bid. The winner then resells the bond, say at 100.5, and if it paid 99, the "underwriters' spread" would be 1.5 ($15 per $1,000 bond). A study by Reuben Kessel of literally thousands of such bids revealed that the spread was, say, $14 when twenty groups bid, and $20 when only one group bid. But if even *two* consortia bid, the $20 fell to about $17, and with three bidders it fell to $16. Even one or two rivals brought the spread surprisingly close to the competitive level. Competition is not easily suppressed even when there are only a few independent firms.

Whether for these or other reasons, more and more economists have come to believe that competition is a tough weed, not a delicate flower. This shift of opinion preceded the recent emergence of powerful international competition for the American market in automobiles, steel, and the like, but this competition may have reinforced the new skepticism of the pervasiveness of monopoly. And accordingly, the enthusiasm for antitrust has diminished, although it has not disappeared. I have begun to appear as an expert witness in a few antitrust cases in recent years, including the attempt by Mobil to take over (buy out) Marathon Oil Company in 1982.

Monopoly

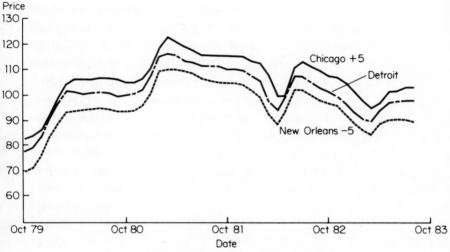

Wholesale gasoline prices in cents per gallon,
October 1979–October 1983 (Platt and Lundberg surveys)

Antitrust cases are often unbelievably long and expensive, but takeover bids are speedy affairs. If the target company opposes the effort, whether to get a better price or to protect the jobs of the officers of the target company, it seeks a court injunction against the takeover; the Mobil-Marathon trial was only one week long. The crux of the case was this: Mobil and Marathon were both substantial wholesale suppliers of gasoline in several midwestern states such as Michigan. If these states individually or collectively were separate markets, the merger *might* impair competition; if they were all part of one vast midwestern market, competition was most unlikely to be reduced by the merger. I testified that the market was much wider than the affected states, and a well-known economist argued the opposite. I based my opinion largely on evidence such as the accompanying chart,

105

which shows the wholesale price of gasoline in Detroit, Chicago, and New Orleans.[5] (Note that Chicago is drawn five cents above its actual price, and New Orleans five cents below, to keep the lines from becoming hopelessly confused.) How could the price possibly move in so similar a fashion in the three widely separated cities if buyers or sellers were not constantly able to shift their trading to where buying prices were low or selling prices were high? Hence the cities were in the same market. The economist for the other side argued that buyers simply could not shift to other cities to make their purchases when prices were less elsewhere, and left the similarity of price movements unexplained.

The judge ruled against Mobil (and me!) and I was aggrieved. My lament may have been due to vanity, but it had an altruistic element. U.S. Steel was able to buy Marathon for almost $1 billion less than what Mobil had offered, and the stockholders of Marathon (I was not one) were the true losers of the case. Is this a purpose of antitrust policy, to allow target firms to select buyers who will treat the officers of the target company better and the stockholders worse? The plain fact is that antitrust policy has often, even increasingly, been bent to such perverse ends, protecting rather than challenging sheltered and inefficient enterprises. Indeed, the antitrust laws have become a hunting license for lawyers, with the tripling of damages the reward for successful suits.

Consider the private antitrust actions. A filling-station operator named Bogosian in the Boston area sued all the major oil refiners on behalf of all filling-station operators.

---

[5] The chart previously appeared in G. J. Stigler and Robert A. Sherwin, "The Extent of the Market," *Journal of Law and Economics* 28 (October 1985): 571.

His complaint was that he was not allowed to sell any gasoline except, say, Texaco in Texaco pumps, and this deprived him of competitive sources of supply. The case, I thought, was utterly destitute of merit. It would be an impossible task for Texaco to monitor the quality of gasoline sold through its pumps, for that quality could vary with every purchase of a tank load of gasoline. Trademarks such as Texaco are literally brands. A trademark allows us to trace a commodity to its maker and therefore allows us to reward or punish that maker accordingly as his product pleases or displeases. An unbranded commodity is often called "generic"; a better name would be "maverick." Texaco would be held responsible for the quality of gasoline sold through its pumps even though it had no control over that quality. Moreover, Bogosian was not compelled by Texaco to have only the pumps of Texaco gasoline at his station. No significant barrier stood in the way of his shifting to another major or minor brand, and no evidence was given of collusion of the refiners in setting the terms of their franchise contracts, which varied substantially from one major refiner to another. So I was convinced of the merits of the refiner/defendants (for one of whom I was a consultant). Yet the refiners threw in a small towel, for fear of the uncertainties of the verdict of an unsophisticated jury. The $30 million towel was divided more or less evenly between a score of lawyers and 200,000 filling-station operators.

This tale could be reproduced many times a year, and it leads me to believe that changes should be made in the law governing private antitrust cases. At a minimum, it would be desirable to compel these class-action plaintiffs to pay the legal costs of the defendants when the latter won. That alone would take much of the fun out of the bringing of cases with small probabilities of success.

The declining importance of monopoly as a problem in public policy or as a hypothesis to explain business behavior is an important reason why my own research interests shifted increasingly to the governmental regulation of economic life—and the reciprocal regulation of government by economic groups.

No End to Means

I have already discussed Gardiner Means's marked influence upon economic thinking during the Great Depression. During much of the 1950s and 1960s he continued to exert a strong influence upon the thinking of the public and the economics profession about how the American economy operates.

Ask a representative sample of economists (or lawyers) the following questions:

1. Do industries dominated by large companies lower prices when business declines? Answer: Usually not.
2. Are the officers of large corporations in control of the corporation's destinies, or are the shareholders in control? Answer: The officers.
3. Is the larger part of the wealth of America possessed by the 200 largest nonfinancial corporations? Answer: Yes.
4. How can we ascertain empirically whether an industry is competitive or monopolistic? Answer: Look at the concentration of control over output. More precisely,

look at its concentration ratio: the share of the industry's sales made by the four largest firms. If this ratio is high (say 75 percent or more) the industry is probably not competitive.

Means gave those answers. The first two views are not original to him, but he was surely the most influential exponent of each of the four. And that is testimony from me, even though I believe that each of the answers is wrong or seriously misleading.

The subject on which Means and I clashed for a decade or more was the first proposition I attributed to Means: that prices are unilaterally set ("administered") by large corporations, and once set, are held at this level for substantial periods regardless of changes in costs of production or changes in the demand for the products. That thesis was first proposed in 1935 in an immensely influential memorandum to the Secretary of Agriculture, Henry Wallace, that was soon published by the United States Senate as "Industrial Prices and their Relative Inflexibility" (Document 13, 74th Congress, 1st session). The severity of the Great Depression of 1932 to 1940, as I have already noted, was blamed by Means on the failure of the price system to work effectively in this new world of administered prices.

My skepticism of this doctrine was shared by numerous economists by the late 1940s, but I was provoked to attack the theory only after Means reformulated it to also explain the appearance of inflation in the 1950s. Now he argued that the giant corporations raised prices simply to increase profits, utilizing a "zone of discretion" in pricing conferred upon them by their large size and market power. This thesis was seized upon by Senator Estes Kefauver as the vehicle for a sustained attack on big business in hearings he held beginning in 1957 (Subcommittee on Antitrust and Monopoly of the Committee on the Judiciary, Part 1). Gardiner Means was the star witness on this occasion, and he presented dramatic charts to show that price increases from 1953 to 1957 were posted only by industries with administered prices; prices, to repeat, set by large companies in cool disregard of the "supply and demand."

I was skeptical of the existence of rigidly set prices, and soon had an opportunity to investigate them. I became chairman of a committee on the price statistics of the federal government (created in 1959 at the request of the then Bureau of the Budget) and the committee made studies of "wholesale" prices, upon which Means's findings were based. One study, by Harry E. McAllister, found that the prices actually paid by buyers were much more flexible—changed much more often—than those reported by sellers to the price-collecting agency (the Bureau of Labor Statistics). This and other evidence led me to conclude that there was no evidence supporting Means's explanation for inflation. Indeed even his exact procedure, if carried on beyond 1957, showed a negative rather than a positive relationship between the rate of increase of a corporation's prices and either the existence of administered prices or the share of an industry's sales made by large companies.

I think that it was about this time that Gardiner Means and I met at a meeting of the Committee for Economic Development. Means approached me and said, "George,

I'm not as stupid as you think." I was flabbergasted and I don't recall what I stammered in reply. I hope I did not then, as I most certainly do not now, think Means was stupid. Indeed, it is extremely improbable that a stupid person could make mistakes so important and successful.

Because the nature of the price system is so important to economists—and to the workings of the economy—I returned to the study of the prices actually paid by business for an important set of prices (of steel, chemicals, nonferrous metals, etc.). I lured James K. Kindahl away from Amherst College, and for two years we went to large businesses and asked them for *buying* prices over a ten-year period. We chose buying prices instead of selling prices because if a seller gives lower prices to some buyers he may be accused of price discrimination under the Robinson-Patman Act, but the lucky buyer who gets a preferred price is essentially free of this legal threat. Our work appeared as *The Behavior of Industrial Prices*, from the National Bureau of Economic Research in 1970. We argued that our evidence was persuasive that actual prices were more responsive to decreases in market demand than the official Bureau of Labor Statistics' prices (which are largely quoted rather than actual transaction prices). Means violently and repeatedly denied this to be the case. Indeed, when a conference commemorating the fiftieth anniversary of *The Modern Corporation* was held at Stanford University in 1982 (see the *Journal of Law and Economics* for June of 1983), Means appeared to defend his original hypotheses with old-time vigor.

It is fair to say that Means's doctrines are much less widely held in economics than they were in the Great

Depression, but their persistence is a remarkable tribute to their palatability to ruling political thought. Once an idea is widely accepted, it is guaranteed a measure of immortality. Its decline in popularity is more often due to changing interests than to contrary evidence, no matter how powerful that evidence may be.

CHAPTER 7

Political Regulation of Economic Life

SUPPOSE that the streets in our town are untidy, strewn with paper, cans, and trash. What can we do about it as individuals, beside taking care not to contribute to the disarray? We could launch a campaign among the people we know and write letters to the local newspaper. But even if we had a measure of success, it would likely be temporary: Campaigns lose their fervor and new issues displace them to vie for our attention.

There would not seem to be any feasible way to deal with this small problem continuously and effectively other than by turning to the town's government for help. The town can pass a law against littering the streets, and punish people who still litter. Better yet, the town can hire people to clean the streets. This is a parochial example of a universal social problem. If individuals or the ordinary associations of individuals (the church, the PTA, etc.) cannot do something, or cannot do it well, where else can one turn but to the state? Unless one is wholly resigned to the world as it is, the state becomes the "Savior in Residence."

The propensity to turn to the state with any problem that cannot readily be solved by individuals and private associations seems to be a natural reaction. The patron saint of private enterprise, Adam Smith, offers a striking example. He lamented that the increasingly more narrow specialization of labor stultified manly feelings and skills, above all the martial spirit, a spirit he believed essential to the preservation of free men. So he asked the state, rather timidly to be sure, to undertake the task of reviving or restoring courage in its citizens. What a task to give to government! A ruthlessly despotic tyranny might destroy men's courage, but what manner of government can instill courage and independence—except, indeed, by practicing self effacement? Yet Smith was no naive admirer of the state. He devoted several hundred pages of his treatise, *The Wealth of Nations*, to the absurdities of the imperial protectionism (called mercantilism) practiced by Great Britain in the eighteenth century.

Smith and his predecessors and successors almost always concentrated on advising the state what it *should* do, or refrain from doing. I certainly did for the first decades of my life as an economist. Hardly ever did anyone undertake the different and more fundamental task of explaining what states actually do, of discovering what are the forces that determine which policies will actually be adopted by a government. And yet, what is the purpose in urging a state to have, say, free trade, as we economists have so vigorously done for two centuries, when protectionism is common and persistent?

The explanation of public policies we have recently learned to give is that they are usually designed by particular groups in a society for their own benefit. That is scarcely news: James Madison found the central task of government to be the control and attenuation of factions (see *The Federalist* No. 10). Yet all he could offer in solution were twin hopes: that democratically chosen representatives would filter out the worst threats to minorities of direct democracy, and that the larger a polity was, the more difficult it would be to form oppressive majorities.

Only in the past three decades has a general theory of the behavior of governments begun to appear. Three scholars, Anthony Downs, James Buchanan, and Gordon Tullock, began the task of constructing such a theory. It is interesting, but not accidental, that economists initiated the theoretical study of the actual, in contrast to the desired, functions of the state. (We also claim Tullock though his graduate degree was in law.) Of all the social scientists, only economists possess a theoretical system to explain social behavior, and it was evidently easier for economists to address unfamiliar political problems with their theoretical apparatus than it was for the political scientists to see the relevance of, and need for, our theory of rational behavior in explaining political phenomena. I started to work in this area shortly before the Buchanan-Tullock book *The Calculus of Consent* appeared in 1962, but my interest at first was much more in the actual effects of economic regulations than in the explanation of why governmental policies are instituted. About this time Claire Friedland joined me.

Claire has now been working with me for almost thirty years, and I acknowledge her roles not only as an invaluable colleague but also as my professional conscience. The computer has made it easy to fish for results. If the statistical analysis doesn't come out "right" the first or the twentieth time, one can drop a year from the data, add a new variable to explain contradictions, take the logarithm of another variable, and so on until, lo, the desired answer appears—all in just a few minutes. I don't know how many times Claire handed me a shattering printout from the computer and said in effect, "Do you still see any merit in this wretched idea?" It has been very difficult to look Claire in the eye and say at this point, "Let's try seventeen more variations on the theme." I must add, with resignation, that she is the most gifted and outrageous punster I have ever met.

We made three studies of the effects of regulatory policies during the 1960s. The first was to determine whether the regulation of electricity rates had an appreciable effect on the rates charged by utilities. The answer: No. (We used comparisons of states with and without regulatory commissions to reach this conclusion.) The common reactions of our fellow economists were either that they knew this already (so why hadn't we saved the work by asking them), or that we were wrong. Later results by Gregg Jarrell suggest that regulation led to *increases* in rates in the early years. The second study sought to discover whether the elaborate review of the prospectuses for new security issues by the Securities and Exchange Commission had benefited the buyers of new issues. Again, our answer was: No. This time our results were greeted with incredulity and violent criticism by econo-

I was a footsore Boy Scout each summer in the Cascade Mountains.

With my parents on the eve of World War II, which my father claimed was "merely a police action," thus avoiding paying me our $10 bet.

Margaret (Chick) and I
in Indiana, Pennsylvania,
shortly before our wedding.
She did not share
her father's fear of
a Hungarian husband.

Jacob Viner, the omniscient historian
of economics and the stern
disciplinarian of price theory, 1948.
(Photo: Department of Economics,
University of Chicago)

Henry Calvert Simons,
the crown prince of
the Chicago School, 1944.
(Photo: Office of News
and Information, University
of Chicago)

Arthur F. Burns as the
brooding eminence of the
Federal Reserve System,
in my photographic version,
1970s.

On sabbatical in Switzerland in 1955 with Chick, Stephen, David, and Joseph (better known as Chop).

As a professor at Columbia, 1951–52, I was no doubt seeking to be as creative as an eraser permits. (Photo: Victor Jorgensen for FORTUNE Magazine)

Wearing our conservative hats at the first meeting of the Mont Pelerin Society are Milton Friedman, GJS, and Aaron Director, in Switzerland, Spring 1947. The Society might better have been called the "Friends of Friedrich Hayek." (Photo: A. Hunold)

A famous if unposed picture of Friedman and me. Note that I am to his right.

Prophet Frank Knight and three of his disciples, Ronald Coase, Stigler, and Harold Demsetz at the University of Chicago, ca. 1965.

Milton Friedman is always young and energetic and it's hard to believe
his friends were celebrating his sixtieth birthday (1972).
LEFT TO RIGHT: Stigler, W. Allen Wallis, Milton Friedman, George P. Shultz,
Herbert Stein, Arthur Burns, and Homer Jones.

The famous Industrial Organization Workshop at the University of Chicago,
likened to a gladiator's pit, in which pride and composure were
often shattered but blood seldom spilled. TOP: John Filer, Rodney Smith,
Rachel McCulloch, Charles Cox. BOTTOM: Reuben Kessel, Marc Nerlove,
(Judge) Richard Posner, ca. 1973.

The King and I: Nobel awards, Stockholm, 10 December 1982.
(Photo: © Svenskt Pressfoto, Stockholm)

GJS, 1982. (Photographer: Michael P. Weinstein)

mists loyal to the regulatory system, but a series of later studies broadly confirmed our results.

The third study asked a question difficult to answer: Have the antitrust laws made the American economy more competitive than it would be in the absence of these laws? One way to approach this question is to make comparisons with other countries that did not have antitrust laws, and we tried this approach. One immediately learns how meager the information about firms and industries abroad is in comparison with American sources. One could get a good statistical history of the production by firms in the steel industry in America in about a week, but to compile an inferior history of the industry in Britain required at least a month. Businessmen abroad are much more reticent. I was guided through a Longine watch factory in 1955, and told by its senior officer at the outset to feel free to ask any questions. A question about profitability so outraged his sense of propriety that most of the rest of the tour was conducted in chilling silence. In any event, the findings of the antitrust study were tentative and more ambiguous, although on balance it appeared that our policy had made the American economy somewhat less prone to effective collusion and had lowered concentration in some industries. These limited results spared us both criticism and compliments.

The most remarkable feature of these studies was not their results but that they were undertaken in the first place. When we made the first serious study of the regulation of electrical rates in 1960, the industry had been under state commission regulation since 1907. Mountains of legal and economic literature contained only one poor study by The Twentieth Century Fund of the effects of regulation. When we tackled the SEC review of new security issues in 1964, that was the first study of the effectiveness of their policy since the commission began thirty years earlier. No such novelty can be claimed for our study of antitrust policy—hundreds of lawyers and economists had passed upon its achievements. Here our claim was only that unlike our predecessors, our program was *improvable* in the sense that the next researcher could start where we left off. For example, we had compared concentration in seven industries in Britain and the United States: Someone with the patience of Job could now add 10 or even 100 more. Fortunately that is now all changed. The number of studies by economists of the effects of regulations exceeds literally a hundred each year. Recently I looked for articles on the effects of usury laws on the volume of residental construction, and immediately found two dozen.

Unlike the miracle of St. Denis (who picked up his head and walked away after being beheaded, leading Voltaire to remark that the first step was the hardest), the second step was harder: Why are some industries and activities regulated by the state, and not others? Farmers, labor unions, manufacturers, symphony orchestras, universities and their professors—all seek and obtain subsidies, suppression of competition, public setting of prices, and the other goodies a state can confer. Other industries and occupations (restaurants, junior business executives, unskilled laborers) fare meagerly at the governmental trough. Why the differences? The answer proves to be complex and elusive, but progress is being made.

One finding has been that small groups do better in politics than large groups, certainly per capita and at times in the aggregate. Today American farmers and their families comprise a little over 2 percent of the population but they get about $40 billion in governmental outlays for income stabilization and all kinds of regulations and special programs (restriction of crops, preferential interest rates, subsidized electrical utilities, etc.). As late as 1950 farmers and their families were about 15 percent of the population and received only about $3 billion from total federal government outlays, as compared with $40 billion (about $9.5 billion in 1950 dollars) in the 1980s. That is about a sevenfold increase in payments per farm operator in dollars of stable purchasing power.

There are two reasons why smaller groups do better than large in the political arena. The smaller group is more cohesive: It is easier to organize the small group, collect funds for lobbying, and keep it informed. There are only about 70,000 beekeepers concentrated in a few western states (yes, there is a federal program for them) but millions of occasional consumers of honey. And secondly, it pays each member of a small group to invest resources in politics, because the payoff will be larger. Each beekeeper gets hundreds of times as much out of the federal program as each taxpayer loses.

Another finding is that no matter how disinterested the goal of public policy, the policy is bent to help politically influential groups at the cost of the less influential. The Clean Air Acts provide striking examples. The basic law has a nondegradation clause: No matter how good or bad the quality of the air in a region, new investments that lower air quality will not be permitted. My colleague, Peter Pashigian, has shown that this policy was instituted to prevent migration of industry from the North Central area to the South and West, and that the pattern of political support and opposition to the policy in Congress clearly revealed these interests. Yet it is inefficient from the national viewpoint to prevent the movement of firms to areas where their emissions would cause little harm. Again, all coal-burning, electrical-generating plants must introduce scrubbers in their chimneys to reduce the emission of sulfur oxides. This clause is designed to prevent the use of low sulfur western coals, for which scrubbers are unnecessary, in order to benefit high sulfur coal from eastern mines. The clause has increased pollution by delaying the construction of new generating plants and inducing the continued use of old plants. The abatement of air pollution, an admirable social goal, is largely thwarted by these special interest policies.

These examples of the economists' findings are representative in their showing that groups possessing political influence use the political process effectively to increase their incomes. That would be an occasion for complaint only from the envious losers if that were the whole story. Of course it is not. When we spend much more than necessary to reduce sulfur oxides in the air, for example, the eastern coal miners and operators gain less than the rest of the nation loses. We could produce the same amount of electricity with the same reduction of undesirable emissions at a substantially lower cost, so the miners' gain is less than the electrical consumers' loss. Yet it is also possible that if we must (for political or other reasons) support Appalachian miners, this is as good a way as any of doing it.

Economists have not only become diligent students of

public regulation, but also they increasingly have served as regulators. My own experience is modest: I was a public member of the Securties Investor Protection Corporation during its initial three years (1970–1973). I was the only economist member, and the remainder of the board was composed of members of the brokerage industry (including Donald Regan, then head of Merrill Lynch) and civil servants. My role was to oppose (sometimes successfully) petty regulatory details (e.g., where brokers should display their SIPC logo) and to make a wholly unsuccessful attempt to vary the assessments of brokers in proportion to the riskiness of their capital structures. The industry members insisted that we not be paid, to avoid the onerous regulations governing federal employees, so I was literally a free rider on their policy.

One could compile a fairly impressive list of economists who have held prominent regulatory positions. Here are a few:

Arthur F. Burns: Chairman of the Federal Reserve Board;

George P. Shultz: Secretary of Labor, Secretary of the Treasury, and Secretary of State;

Darius Gaskins: Chairman of the Interstate Commerce Commission;

James C. Miller III: Chairman of the Federal Trade Commission and Director of the Office of Management and Budget; and

Alfred Kahn: Chairman of the Civil Aeronautics Board.

What is interesting about the list, is that each of these

economists except Arthur Burns was a strong proponent of deregulation and free markets: Burns supported wage and price controls. I was unprepared to find so strong a record in favor of free markets; after all, Adam Smith spent his declining years as a Commissioner of Customs for Scotland, and showed great diligence in the pursuit of smugglers.

CHAPTER 8

The Economist as Expert

ECONOMISTS ARE EXPERTS, at least in the eyes of legislators and judges and juries. That is not high praise; it means that economists have professional training, but not necessarily that they possess a body of useful knowledge that will be almost universally accepted by similarly trained people. It is notorious that one can find psychiatrists to testify on either side of a question: Was X insane, or at least incompetent, when he shot the President? It is not difficult to find some professors of economics from major universities to testify that industry Y is competitive, and others to testify that it is monopolized. An expert is simply someone who should know more than an intelligent layman about the problem in hand.

Expert testimony to Congress by academic economists is usually objective in one narrow sense: No one pays the economist to do it. I must confess to my annoyance at the importance assigned to this fact by almost everyone.

Being paid to do something is the basis of the distinction between an amateur and a professional, and professionals are commonly more skilled than amateurs. Yet in connection with pay for testimony most people think instead of the difference between a prostitute and a virtuous woman, and therefore of motive or conduct rather than skill or knowledge. In complex matters of public policy one would think that knowledge was more germane than motive; the public can supply its own motives. Indeed one cannot pose as an expert unless he is paid to be one: The respected professor won a coveted appointment at a good university; no one would listen to him if he were an unpaid volunteer from a lesser college.

In any case, the economist appearing before a congressional committee is there to give advice on economic policies. I am not an authority on such appearances: I have made only three, and one was compelled by the issuance of a subpoena. My first appearance was in 1950 (my testimony on competition in the steel industry is sketched in chapter 6). Both Edward Levi and chairman Emanuel Celler were apparently pleased with my vigorous, not to say outrageous, complaints against the steel industry, and I was invited back at the end of the hearings to summarize the testimony on that industry. This was perhaps the first of numerous times that I was told that I would not be paid for my appearances in order to preserve my—well, it couldn't have been integrity in this case if the Congress had paid me, so it must have been the appearance of integrity.

My second appearance was less controversial. A Price Statistics Review Committee had been formed in 1960 by the National Bureau of Economic Research, and it reviewed all the major price-collecting programs of the federal government for the (then) Bureau of the Budget. Thus it examined the Consumer Price Index (the so-called cost of living) to see how the Bureau of Labor Statistics treated changes in the quality of goods (mostly by neglect), changes in retail outlets (belatedly), changing family-buying patterns (once a decade or so), and so on. The committee presented its report to a polite Joint Committee on the Economic Report, chaired that day by Senator Paul H. Douglas. It recommended a large expansion of the price-reporting program, which did not arouse the opposition of the price-collecting agencies.

That report illustrates the favored position of the academic world. When a private industry wishes the government to do it some favor, it has to curry favor with a legislator or two and hire expensive lobbyists. Here professors were lobbying for better economic statistics, and the federal government paid for the investigation, admittedly at very modest rates. The publicly acknowledged benevolence of academic institutions and personnel is a source of wonder to me. The public's attitude is illustrated by the fact that a federal judge may teach at a university, but is denied other forms of nonjudicial employment. Could this attitude have survived from the time when the chief function of colleges was to train young clergy? The attitude has survived the obvious self-serving eagerness of the physical scientists to spend half of the nation's income if given the chance. The social scientists would settle for what the physical scientists are already getting, thus displaying proportionate greed.

My last appearance before a congressional committee was involuntary. I had been chairman of a task force on

Productivity and Competition to advise incoming President Nixon in early 1969 on the antitrust policies the new administration should pursue. Our report received a hostile reception from the new Attorney General, John Mitchell, who refused to publish it. The main objection, I conjecture, was that we stated that the large conglomerate mergers were usually innocuous in their effects on competition.[1] Mitchell soon launched an attack on conglomerate mergers, I believe because he thought it would show that the Republicans were not soft on big business. Of course our report was soon leaked, probably by a sympathetic staff member in the Antitrust Division of the Department of Justice.

In due time the Subcommittee on Small Business and the Robinson-Patman Act of the House Select Committee on Small Business scheduled hearings on my task force's report and summoned me to appear. I saw no reason to comply. As a private economist I had performed a service for the incoming president and saw no need to appear before a hostile committee. I turned to two distinguished lawyers, who were also summoned, and asked them if they would defend my right to stay home. They informed me that they were not going to jail and showed no inclination to establish new legal rights for me. So when the U.S. Marshall, Mr. John D. Miszner, relayed the command from the Honorable John D. Dingell to appear "without fail" before the subcommittee, I did not fail.

The issue was our task force's treatment of the

---

[1] Mitchell, in meeting with our task force, focused upon the question: How could the antitrust laws be used to combat organized crime? I consider it a weakness of character that I answered courteously if negatively instead of lecturing him on his ignorance.

Robinson-Patman Act, an act that forbids price discrimination (different prices of the same commodity for different classes of customers) where the effect may be to impair competition. The law sounds innocuous but it has been a powerful tool to combat new forms of competition (e.g., the chain stores) and to harass competitive practices. I opened my appearance with a brief statement, of which the following part is the essence:

> Our report devoted little space to the Robinson-Patman Act, but that little attention was highly critical. Other members of the Task Force may have had other reasons for adopting this position; mine is as follows.
>
> Price discrimination that is genuine, with a price structure substantially indefensible on cost grounds, is evidence of monopoly and monopoly is an inefficient and objectionable form of economic organization. The Robinson-Patman Act is right in opposing price discrimination, but errs in its method of attack:
>
> 1. It defines price discrimination improperly, and in particular identifies it with price differences, so it attacks desirable price structures as well as undesirable ones.
> 2. It attacks one expression of monopoly power but leaves the monopoly power itself undisturbed, and still operative.
> 3. It fosters collusive and anti-competitive practices, and protects obsolete and inefficient types of business.
>
> Primarily for these reasons the Robinson-Patman Act is opposed by virtually all economists. I hope the Subcommittee will reflect upon the fact that if all the prom-

inent economists in favor of the Robinson-Patman Act were put in a Volkswagen, there would still be room for a portly chauffeur.[2]

A small gauntlet, to be sure.

The appearance before the committee was less interesting than the negotiations over my appearance. The chief counsel repeatedly pressed the question: Isn't it true that big businesses drive out small businesses by predatory tactics and isn't, therefore, protective legislation such as the Robinson-Patman Act necessary? Just as repeatedly I denied that predatory competition was a common practice, or one that required any legislative control not already in the Sherman Act. The exchanges were not violent:

MR. POTVIN: Now, as you point out in your book *The Organization of Industry*, sir—

MR. STIGLER: I am delighted with your choice of reading matter.

MR. POTVIN: I won't bring out how many hours of sleep this has cost trying to understand some of your more rigorous and elegant models. On page forty-four, sir, you say that: "It follows that oligopolistic collusion will often be effective against small buyers even when it is ineffective against large buyers. . . . Let us henceforth exclude small buyers from consideration." Of course . . . one of the goals

[2] George J. Stigler, Hearings, Subcommittee on Small Business and the Robinson-Patman Act of the House Select Committee on Small Business, October 8, 1969 (Washington, D.C.: U.S. Government Printing Office, 1969).

> ... is that the small businessman ... would not be
> entirely "excluded from consideration".
>
> MR. STIGLER: You are obviously joking. It is all right. The
> exclusion was for the purpose of discussing one
> subject at a time.
>
> MR. POTVIN: Of course, Professor. But I think that in a
> less jocular vein; if I may, you make a point here of
> the first magnitude of importance.[3]

Clearly the forty-eight pages of interrogations that Richard Posner, my fellow witness, and I underwent that day had no purpose except to publicize the opinions of the Small Business Subcommittee.

Indeed a legislative committee should not be influenced in any important degree by expert testimony, however learned it may be. Legislation should be based upon tested knowledge, and the hearings before such committees is no way to test knowledge or opinions. Science tests knowledge by having different scholars seeking to replicate the same results, and by collecting new evidence to test theories. The hearings can do neither, partly because they do not provide the time for careful scrutiny of testimony, and partly because the experts themselves are chosen by interested parties.

The setting is quite different when the economist appears in litigation because the legal system acts upon the adversary principle, namely, that the vigorous presentation of the merits of each side of an argument by the supporters of the respective sides is the best way to deter-

[3] George J. Stigler, Hearings, ibid., note 2.

mine the facts in dispute. Hence everyone except judge and jury are appearing as proponents of particular views.

Let me begin with two true stories of economists who gave expert testimony. The first involves my University of Chicago professor, Jacob Viner, who in about 1925 testified as an expert for the government in an antitrust investigation involving the steel industry. As Viner recounted his participation to me many years later, he began testifying as an objective academic student of price theory. After sharp cross-examination by the attorneys for the other side, however, he found himself becoming more and more defensive of the government's position and more and more critical of the steel companies. The adversarial process had turned him into an adversary.[4]

The second story involves a friend who had been deposed preparatory to a private antitrust suit (one between two or more private parties). The deposition is the instrument by which each party learns what the testimony of the other party's experts will be, so they will be prepared to deal with that testimony in the actual courtroom trial. My friend gave me a copy of the transcript of his deposition and asked: "Am I still an honest man?" I read his deposition and assured him that he was still an honest man. "But," I said, "thank God we don't talk like that in the university." He had said nothing that was remotely untrue, but had left unsaid a good deal that was true. I do not recall the details, so I'll make up an entirely fictitious illustration.

[4] For a related episode involving Viner, see his review, "Objective Tests of Competitive Price Applied to the Cement Industry," *Journal of Political Economy* 33 (1925): 107–11, where he takes to task H. Parker Willis and J. R. B. Byers, *Portland Cement Prices* (New York: Ronald Press, 1924), in which the competitiveness of cement prices is defended.

Examination by the Expert's Side

In the Courtroom	In the University
Q: The defendants all charged the same prices for the product?	
A: Yes.	A: Their list prices were the same, but the prices actually charged differed to an unknown degree, but probably substantially.
Q: Is price identity found in industries where the firms collude in setting prices?	
A: Yes.	A: Yes, and in all other industries where buyers have good information and the products of the firms are essentially identical.

Cross-Examination by the Other Side

In the Courtroom	In the University
Q: Did not the various firms actually charge different prices at a given time?	
A: There were no doubt episodes of that sort but I found no evidence to suggest that price differences were important.	A: My attorneys said we didn't have time or resources to investigate actual prices.

Q: Were the accounting profits of the defendant companies unusually large?

A: Not the reported profits.	A: No.

Q: Is this not evidence of competitive pricing?

A: No. Accounting profits are subject to many influences and accounting practices that make them unreliable for competitive analysis.	A: The evidence is not conclusive: One must look carefully at how assets are valued, depreciation charges set, and overhead costs allocated.

A cardinal rule of expert testimony is not to volunteer any answers. Another rule, for lawyers, I am told is: Don't ask a question if you don't know what the answer will be. Spontaneity is anathema to the legal process.

After more than forty professional years during which I was never a courtroom expert, I became involved in a number of antitrust cases. I find it difficult now to identify the motives for many things I have done, and the same is true in this case. An expert is well paid but that cannot be the whole story since it would have been true at an earlier time. I was curious to see how economics operated in a courtroom, and I like the challenge of the adversary process.

I was curious also as to how I would behave. Professors—perhaps all nonprofessors also—love to parade their virtue. When I was at Columbia University, the younger Van Doren revealed his connivance in a dishonest TV information quiz program in which he had won

$140,000. I did not join the chorus of righteous disapproval: I was not so confident that the run of professors would resist large temptation (the prize was over $500,000 in 1988 dollars). I do not have a good answer to the question of my own behavior after the dozen or so legal cases I've appeared in, and not just because a man should not judge his own case. I do know that Viner was correct when he predicted that one becomes engaged. In my case, unlike Viner's, I have seldom been brought to dislike the experts or lawyers on the other side, but I have usually ended up liking those on my side, which comes to much the same thing. That has not led to a conscious manipulation of the evidence and arguments but to a cast of mind that focuses on the arguments favorable to one's side. Is the expert honest? At very best, probably as honest as is possible in a process in which truth is sought by the vigorous presentations of opposing views, and where any admission by one side is heavily overemphasized by the other side.

And that ambiguous answer applies only to the most virtuous of experts. I have been shocked by the testimony of economists in many cases. Consider the 1948 Cement Case (*Federal Trade Commission* v. *Aetna Cement*) in which one economist testified that it was perfectly possible for ten producers to submit sealed bids that were identical to ten decimal places in the price per barrel of cement, without prior agreement. The delivery of cement was to be well after the date of the bidding. It is simply inconceivable that such identical bids could be achieved without agreement. That case took place decades before I became a witness, but I have encountered the same astonishing flexibility more recently. In one case, for example, the plaintiff had been convicted of commercial bribery, that is, giving secret bonuses to purchasing agents of companies who bought from the plaintiff. This plaintiff's economist was quite unwilling to condemn this practice, yet bribery subverts the purchasing agent, who sacrifices his employer's interests for his own gain.

Of course the partisanship of experts is partly a matter of selection; an expert is not put on the stand when the party who has employed him does not like his testimony. I was once asked in a famous trial to testify that if one large company was forced to sell its shares in another large company, the price of the shares of the latter company would fall precipitously. When I replied that the value of shares is governed by expected earnings, not the remarks of federal judges, the temperature in the room fell precipitously. Of course, I was not hired. The economist who was hired dealt forthrightly with the question: He said he had not been able to prepare on that matter. Diogenes would not have accepted that economist, and I doubt whether any expert witness would be approved by Diogenes. Everyone in a legal proceeding, except the jury, knows that this is how experts behave, so no one—the jury aside—is misled. And it may be that on balance the jury is aided in reaching its decision by the polarized presentations of the alternatives.

The economist's engagement with policy usually becomes a full-time activity when he enters government service, whether as a lower level technician or as a lordly member of the cabinet. It is interesting that in twenty years prior to 1982, a total of thirteen professional economists had reached cabinet rank (see the examples on page 121), and even more interesting is that this practice reached its peak in the reign of Richard Nixon, who was scarcely the favorite of the academic world.

The very nature of the political process demands as much loyalty to the administration by the economist as he shows to his side in adversary proceedings. Indeed, political competition is an adversary process, and loyalty to the administration is considered essential to the effective conduct of government. A professional economist's disagreement with the policy of his administration must not reach the public, but when the economist carries his criticisms to the public, he is considered to be an irresponsible subordinate, who improperly chose public dissent rather than resignation. That kind of economist need not expect in the future any responsible position in government from either party.

Indeed a political career must eventually lead an economist at very best to a strained and artificial position. He must be silent on, or at most cautiously critical of, policies he deplores, such as the minimum wage law or acts of protectionism. He must support policies which are not so much wrong as ridiculous. At one time both the chairman of the Council of Economic Advisors, Herbert Stein, and the chairman of the Federal Reserve Board, Arthur Burns, advised Americans to eat cheese instead of meat (which was in short supply) to fight inflation. In short, the economist must accept the basic credo of the politician: In order to do some things that are good, it is essential to accept many things that are bad.

I conclude—and perhaps I am alone in concluding—that when the economist goes to Washington, he deserves no more credence, and no less, than any other political appointment, and it is mildly deceptive to address him as Doctor or Professor.

My own Washington appointments are here briefly recalled. As I mentioned earlier, during 1940 to 1942, I worked, at first part-time, for Leon Henderson, in the group called the Defense Finance Unit, the precursor of the Office of Price Administration. It had an excellent leader, Raymond Goldsmith, and able colleagues such as Herbert Stein and William Vickrey, and we enthusiastically—and of course unsuccessfully—opposed price controls before the unit's demise. An even earlier episode (1935–1936) in the National Resources Planning Board has also been noted. From 1943 to 1945 I was in the Statistical Research Group at Columbia University, working under Allen Wallis and with Milton Friedman, Leonard Savage, and others on applications of statistical reasoning to war-related activities. There we were essentially outside of the political process. From 1969 to 1970 I served on the Blue Ribbon Defense Panel, which had been created by President Nixon to review the organizational structure and operation of the Department of Defense. Gilbert W. Fitzhugh (the head of Metropolitan Life) was its chairman. Members were sought who were "generally unfamiliar with the operations of the department," a condition I certainly met in spades. We were allowed only one year for this task, and the panel was overcome by the magnitude of trying to deal with an organization of such stupendous size and complexity and so riddled with entrenched baronies. Finally, from 1970 to 1973 I was a member of the Securities Investor Protection Corporation, created to cover the losses of customers when brokers failed or embezzled funds. In none of these roles was I ever approached by a newsman or a politician (except in the confirmation hearings for SIPC) so I was never tempted to say anything interesting, true, or false.

I was offered one position that was exceptional: that of foreign trade advisor to the president. George Shultz had persuaded President Nixon to offer the post to me, but to the President's relief I declined. I still am not certain why I declined it. Not because I lacked foreign trade expertise, although that was true. Not because I feared the Ehrlichman-Haldeman group, although Arthur Burns, possessing hard experience, warned me about them. At the time I said it was because I didn't want to go to Washington for at least two years, and that was part of the reason. Quite possibly, however, it was a subconscious recognition that I have an unruly tongue. When I appeared at the White House after being awarded the Nobel Prize in 1982, I caused consternation at an impromptu press conference by talking about the current depression, a word that is apparently obscene in Washington. I was removed from the platform in a manner reminiscent of vaudeville days, which is surely appropriate in a theatrical town. If I had worked in the White House, I assume that I would have learned to be discreet and possibly also to be illegal—to the great benefit of a then undreamed-of autobiography.

CHAPTER 9

The Apprentice Conservative

BEFORE I went to the University of Chicago I suppose I had vaguely liberal political inclinations, but no strong convictions. I also had a most modest knowledge of political and economic life, but my lifetime as an observer of young adults in college convinces me that a modest knowledge is all that is needed for, or possibly even compatible with, strong political views.

At Chicago we students were taught some elements of what most people would call a conservative economic and political philosophy. We began to learn that a competitive economic system performs important economic tasks efficiently. Those goods get produced that people want the most; workers and capitalists are induced to put their labor and capital into the most productive uses; and historically an immense creativity was unleashed. At first these were textbook lessons to be glibly repeated but over many years they turned into a deep and admiring understanding of the extraordinary efficiency and hardiness of a competitive system.

Our Chicago professors were equally dedicated to personal freedom, and they were outraged by the tyrannies of Europe. In 1936 Frank Knight spurned an honorary degree from Heidelberg because of the Germans' treatment of Jews. That hostility to collectivist restrictions on personal freedom, as well as the liking for a competitive order, were somewhat stronger in the University of Chicago's economics department than at most other places (except, and especially, the London School of Economics).

But that was the extent of our brainwashing. As I will note in the following chapter, the attitudes of the Chicago faculty toward economic policy were diverse and often ambiguous. It was difficult for a graduate student to come away from these men with a definite, let alone a dogmatic, plan of economic salvation in the middle of the Great Depression. Henry Simons was perhaps the most influential upon his students with his lucid blueprint of a good society.

The economists at other universities were on average less critical toward the New Deal programs that Roosevelt launched. They did not share so quickly or fully the Chicago economists' outrage at the honeycomb of restrictions imposed by the National Recovery Administration. Economists elsewhere were not on average so incensed as we were by the farm programs. But these other economists were on balance less favorable in the early 1930s to the monetary expansion and public works programs urged by the Chicago economists to encourage business recovery.

Gradually in the later thirties the distinctiveness of the Chicago economists (and their departing graduate students) became more marked as the Keynesian theory began to conquer the young economists elsewhere. But the war soon took most people away from these issues.

Professor Friedrich Hayek, a gifted young Viennese economist, came to the London School of Economics in 1931 and immediately became a major force. He led the attack on Keynes's first and unsuccessful attempt to reconstruct economics, *A Treatise on Money* published in 1930. He revived the Austrian theory of capital and extended it to explain business cycles and he defended it with skill. By the late 1930s, however, his interest began to shift from technical economic analysis to the role of intellectuals in society and to the nature and viability of a free society. Near the end of World War II, in 1944, Hayek wrote a small book, *The Road to Serfdom*. In it he argued that the western democracies were proceeding down the same road that fascist Germany and Italy and communist Russia had already taken, and that that road led inevitably to the loss of individual freedom. The book had been rejected by numerous publishers but was finally accepted by the University of Chicago Press after being strongly supported by Aaron Director. It proved to be a considerable success; over the years the American edition has sold about 210,000 copies. I have recently reread it, and I simply cannot understand why it became popular. I mean this as a compliment to Hayek. He presented a reasoned, lucid argument, but it is an argument largely in abstract terms. There are no dramatic factual claims, no attacks on the motives or competence of opponents, no studied exaggeration of the vices of centrally planned economic systems. Hayek has always been both a gentleman and a scholar. The title contains the book's drama, and perhaps that was enough.

Hayek's style and pace may be suggested by his discussion of one important step of his argument: the thesis that separate governmental interventions in various industries are inherently inconsistent and unstable and must lead to comprehensive, centralized control of the economy by the government.

It is important clearly to see the causes of this admitted ineffectiveness of parliaments when it comes to a detailed administration of the economic affairs of a nation. The fault is neither with the individual representatives nor with parliamentary institutions as such but with the contradictions inherent in the task with which they are charged. They are not asked to act where they can agree, but to produce agreement on everything—the whole direction of the resources of the nation. For such a task the system of majority decision is, however, not suited. Majorities will be found where it is a choice between limited alternatives; but it is a superstition to believe that there must be a majority view on everything. There is no reason why there should be a majority in favor of any one of the different possible courses of positive action if their number is legion. Every member of the legislative assembly might prefer some particular plan for the direction of economic activity to no plan, yet no one plan may appear preferable to a majority to no plan at all.

Nor can a coherent plan be achieved by breaking it up into parts and voting on particular issues. A democratic assembly voting and amending a comprehensive economic plan clause by clause, as it deliberates on an ordinary bill, makes nonsense. An economic plan, to deserve the name, must have a unitary conception. Even if a parliament could, proceeding step by step, agree on some scheme, it would certainly in the end satisfy nobody. A complex whole in which all the parts must be most carefully adjusted to each other cannot be achieved through a compromise between conflicting views. To draw up an economic plan in this fashion is even less possible than, for example, successfully to plan a military campaign by democratic procedure.[1]

The argument is consistently carried at this high level of generalization.

The popularity of Hayek's book led a conservative midwest foundation, the Volker Fund (which is no longer extant), to contribute to the support of a meeting he called in Switzerland for ten days in the spring of 1946. I had never met Hayek but my Chicago teachers certified my eligibility for the coming totalitarian firing squads. It showed my lack of inner conviction of the imminence of totalitarianism that the thought never entered my mind.

It was a revealing first visit for the younger participants, including Milton Friedman and me. En route we

---

[1] Friedrich A. Hayek, *The Road to Serfdom* (Chicago: Univ. of Chicago Press, 1944), 64.

were depressed as much by the austerity of the British economy as by their food (if an ersatz sausage is indeed food). We were instructed as well as embarrassed by the casualness of French life: We did not learn until we left France that we required food ration tickets. I concluded that the British obeyed all laws, the French none, and the Americans obeyed those laws that deserved obedience—in retrospect, something of a simplification. Indeed the black market was a boon to French economic life; it allowed prices to perform their functions. I was instrumental, for the only time in my life, in instructing Friedman on monetary affairs. We sought to convert some dollars into francs at the unofficial exchange rate rather than the official rate that greatly overvalued the franc. I undertook the exchange and approached the clerk at the Grand Hotel, where we were staying. "Could you direct me to the closest outlet for the black market in currency?" I asked. "Go no further, gentlemen" was the response as he extracted a wallet from his jacket.

The thirty-six participants who assembled on April 1, 1947 at the Mt. Pélèrin Hotel, located on a hill with the same name overlooking Vevey, were cosmopolitan and interdisciplinary. They included historians (C. V. Wedgwood and Erich Eyck); philosophers and theologians (Hans Barth, Karl Popper, and Michael Polanyi); a few journalists (Trygve Hoff, John Davenport, and Felix Morley); and a host of economists, including Walter Eucken of Freiburg, Maurice Allais of Paris, Lionel Robbins and John Jewkes of Great Britain, Carl Iverson of Copenhagen, and many Americans. Among those unable to attend, alas, were Luigi Einaudi, who would soon become president of Italy; Walter Lippmann; and my employer, Henry Wriston, the president of Brown University.

In his opening address, Hayek emphasized the role of the intellectual in providing guidance to the western world:

The basic conviction which has guided me in my efforts [to bring this meeting about] is that if the ideals which I believe unite us, and for which, in spite of abuse of the word, there is still no better name than liberal, are to have any chance of revival, a great intellectual task is in the first instance required before we can successfully meet the errors which govern the world today. This task involves both purging traditional liberal theory of certain accidental accretions which have become attached to it in the course of time, and facing up to certain real problems which an oversimplified liberalism has shirked or which have become apparent only since it had become a somewhat stationary and rigid creed.

The immediate purpose of this conference is, of course, to provide an opportunity for a comparatively small group of those who in different parts of the world are striving for the same ideals, to get personally acquainted, to profit from each other's experiences and perhaps also to give encouragement to each other. I am confident that at the end of these ten days you will agree that this meeting will have been well worth while if it has achieved no more than this. But I rather hope that this experiment in collaboration will prove so successful that we shall want to continue it in one form or another.[2]

And continue it we have. I follow Darwin in accepting survivability as the test of an institution, so at least *part* of modern society wished a repeated gathering of old-fashioned liberals. I confess that none of the later meetings I attended equalled for me the interest of the first session.

The discussions were at a high level, and they were not by any means a harmonious chorus. The protection of agriculture and of agricultural classes generally had strong supporters and opponents. The gold standard was the cherished goal of the older members, but not of the younger economists. On the last day Hayek proposed a set of basic principles, not as a doctrinaire creed but as a common ground. The first was that we believed in the dignity and cherished the freedom of individuals. Everyone agreed. The second was that we believed in the institution of private property—and a viper in our midst protested! But he really wasn't a viper: Maurice Allais was a gifted engineer and economist, and at the time he believed that private ownership of land was untenable. (The reason need not occupy us; it turned on the fact that if the interest rate went to zero, as he feared it would, land would become infinitely valuable. Allais subsequently abandoned his capital theory and this fear.)

[2] Friedrich A. Hayek, "Aims and Organization of the Conference," address to the Mt. Pelerin Society, Vevey, Switzerland, April 1, 1947 (mimeo), subsequently published as "Opening Address to a Conference at Mont Pélèrin," in F. A. Hayek, *Studies in Philosophy, Politics and Economics* (Chicago: University of Chicago Press, 1967), 148–59.

My memory has a strange way of selecting its contents. I do not remember much of the ten days of discussion, but I remember the delight with which Walter Eucken (a distinguished German economist who had opposed the Nazis and yet remained in Germany) ate his first orange in five years. I do not remember much of what Frank Knight said but I remember his hailing me as a modern "Fremont the Pathfinder" because I did not get us lost on a short stroll in Vevey.

Before the meeting was adjourned we formed the Mt. Pelerin Society, named after the hill overlooking Lake Leman where we met. The society has persisted to this day, meeting biennially, or more often, over much of the western world and occasionally in Asia. It publishes no books or periodicals, engages in no political activity, issues no proclamations, and its members continue to embrace a fairly wide spectrum of attitudes on economic and political affairs.

I believed much more in the central theme of *The Road to Serfdom* when it first appeared than I do now. The reason is that its main prediction, if true, lies in the uncertain future. It is a fair reading of *The Road to Serfdom* to say that forty years more of the march toward socialism would lead to major losses of the political and economic freedom of individuals. Yet in those forty years we have seen that continuous expansion of the state in Sweden and England, even in Canada and the United States, without consequences for personal freedom so dire as those he predicted.

It is unquestionably true that the American citizen of today is regulated much more than he was when Hayek wrote: Taxes take a substantially larger share of his income and his conditions of employment (licensing, affirmative action, age of retirement, etc.) are extensively prescribed by law. But his range of economic choices has become wider with the spread of higher education and rise of real incomes, and his main political rights (as enumerated in the first ten amendments to our Constitution) have not been seriously impaired.

The proximate reasons that the darkly pessimistic predictions of conservatives have not been fulfilled are two. First, the predictions were based on their special view of freedom: Freedom as consisting only of the lack of coercion by the state, so that the widening range of choices due to the growth of income and education is not an effective increase in freedom in Hayek's view, although it is in mine. The second reason is more interesting. Hayek denied that piecemeal regulations of a hundred different industries and callings could survive. The conflicts and inconsistencies would force the adoption of a single, centralized all-comprehensive plan—and that plan could allow little individual choice. But that multitude of inconsistent, partial interventions by the state in economic life is exactly what we have. Hayek's orderly mind could not comprehend the survivability of our disorderly world.

CHAPTER 10

The Chicago School

THERE WAS NO Chicago School of Economics when the Mt. Pelerin Society first met at the end of World War II. In the 1930s economics appeared to be a little different at the University of Chicago than elsewhere, but the same statement could be made about most major universities.

Frank Knight was skeptical of the moral and intellectual content of political behavior and particularly hostile to central economic planning, but he was also severely critical of the ethical basis of a competitive economy. No doctrinaire defender of private enterprise would find him a source of strength.

Henry Simons had preached a form of laissez-faire in his famous 1934 pamphlet *A Positive Program for Laissez Faire*, but what a form! He proposed nationalization of basic industries such as telephones and railroads because regulation had worked poorly. (I am reminded of the king who was asked to award a prize to the better of two minstrels. After hearing the first, he said, "Give the prize to the second.") Simons urged an extremely egalitarian policy in the taxation of income and detailed regulation of business practices such as advertising. Much of his program was almost as harmonious with socialism as with private-enterprise capitalism. Yet in the area of monetary policy he exerted a strong influence upon the later Chicago School with his trenchant case for a monetary policy conducted according to rule rather than to discretionary manipulation. In particular he urged that the rule be the stabilization of a comprehensive index of prices. This rule is clearly the parent of the later proposal that the money supply should grow at a steady rate, say 3 or 4 percent a year.

Jacob Viner, the other major figure, had nineteenth-century liberal tastes, but rebelled against doctrinaire or simplified or "extreme" positions. The rest of the faculty were highly varied in their policy preferences: Paul Douglas favored a large economic role for the state; Simeon Leland was a traditionalist in taxation; Harry Millis was an old-fashioned labor economist; Lloyd Mints wrote only on central bank policy; Henry Schultz stuck to his mathematical and statistical knitting; and Oskar Lange was a socialist.

Professor Viner and his students of the time such as Martin Bronfenbrenner have testified that they had not encountered the name or the belief that there was a distinctive Chicago School during this early period, and I have found no hints of such a belief in the economics profession before about 1950, and no widespread recognition of the school for another five years.

By the 1960s, however, the profession had widely

agreed that there was a Chicago School of Economics. Edward H. Chamberlin had written a chapter on the Chicago School in his *Toward a More General Theory of Value* in 1957, the earliest such explicit essay I have found. He found the school to be distinguished "by the zeal with which the theory of monopolistic competition has been attacked," and called it the Chicago School of Anti-Monopolistic Competition. What was a minor recreational activity for us was the raison d'être to him! H. Lawrence Miller wrote perhaps the first comprehensive article on the school and its central views in 1962 in the *Journal of Political Economy* and by then the school was treated as well established and widely recognized—and widely denounced.

The origin of the school can be identified only if the central theses of the school are known. They were two: a policy position and a method of studying economics. The policy position was the more commonly recognized element of the school, and clearly Milton Friedman, who had returned to Chicago in 1946, was the primary architect of these policy views. Before that time he had written little on economic policy: some wartime essays on inflation and a pamphlet, *Roofs or Ceilings?*, written with me at the University of Minnesota in 1946 denouncing rent controls.

The situation quickly changed after his arrival at Chicago. Friedman proceeded to establish three lines of work, which together constituted his fundamental contributions to the formation of the Chicago School. First, he revived the study of monetary economics, which had become moribund. He used the quantity theory of money, and refurbished and extended it, not only to study economic behavior but also to launch a powerful attack on the Keynesian School. Second, he presented strong defenses of laissez-faire policies, and invented important new policy proposals. And finally, he developed and employed modern price theory in important ways.

As part of the first task, Friedman concentrated upon the refutation of two central theses of Keynesian economics: (1) fiscal policy—government taxing and spending—was the primary tool of government to influence the level of employment and money income of an economy, and (2) monetary policy—the actions of the Federal Reserve System in our country in fixing the available amount of reserves of banks—was accommodative to economic conditions rather than controlling. A strict monetary policy might hold down economic activity but an expansive policy could not lead to a revival of business. The fashionable metaphor was that one can pull on a string but not push it.

I believe that it is fair to say that Friedman succeeded in dispelling these dogmas, against the opposition of most of the macroeconomists in the United States and England. He established as a powerful empirical rule the proposition that all large changes in the supply of money are associated with large corresponding changes in a nation's money income. The large role of the behavior of money in modern policy discussion—M1, M2, and other concepts frequent the daily newspapers—is surely due more to Friedman than to any other two or five people. He later reminisced on the setting for this work:

I was myself first strongly impressed with the importance of the Chicago tradition during a debate on

Keynes between Abba P. Lerner and myself before a student-faculty seminar at the University of Chicago sometime in the late 1940s (or perhaps early 1950s). Lerner and I were graduate students during the early 1930s, pre–*General Theory*; we have a somewhat similar Talmudic cast of mind and a similar willingness to follow our analysis to its logical conclusion. These have led us to agree on a large number of issues—from flexible exchange rates to the volunteer army. Yet we were affected very differently by the Keynesian Revolution —Lerner becoming an enthusiastic convert and one of the most effective expositors and interpreters of Keynes, I remaining largely unaffected and if anything somewhat hostile.

During the course of the debate, the explanation became crystal clear. Lerner was trained at the London School of Economics, where the dominant view was that the depression was an inevitable result of the prior boom, that it was deepened by the attempts to prevent prices and wages from falling and firms from going bankrupt, that the monetary authorities had brought on the depression by inflationary policies before the crash and had prolonged it by "easy money" policies thereafter; that the only sound policy was to let the depression run its course, bring down money costs, and eliminate weak and unsound firms.

By contrast with this dismal picture, the news seeping out of Cambridge (England) about Keynes's interpretation of the depression and of the right policy to cure it must have come like a flash of light on a dark night. It offered a far less hopeless diagnosis of the disease. More important, it offered a more immediate, less painful, and more effective cure in the form of budget deficits. It is easy to see how a young, vigorous, and generous mind would have been attracted to it. . . .

The intellectual climate at Chicago had been wholly different. My teachers regarded the depression as largely the product of misguided governmental policy —or at least as greatly intensified by such policies. They blamed the monetary and fiscal authorities for permitting banks to fail and the quantity of deposits to decline. Far from preaching the need to let deflation and bankruptcy run their course, they issued repeated pronunciamentos calling for governmental action to stem the deflation—as J. Ronnie Davis put it, "Frank H. Knight, Henry Simons, Jacob Viner, and their Chicago colleagues argued throughout the early 1930s for the use of large and continuous deficit budgets to combat the mass unemployment and deflation of the times."

They recommended also "that the Federal Reserve banks systematically pursue open-market operations with the double aim of facilitating necessary government financing and increasing the liquidity of the banking structure." There was nothing in these views to repel a student; or to make Keynes attractive. On the contrary, so far as policy was concerned, Keynes had nothing to offer those of us who had sat at the feet of Simons, Mints, Knight, and Viner.[1]

It has been disputed whether Friedman faithfully portrayed the earlier Chicago work—most of which was nec-

---

[1] Milton Friedman, "Comments on the Critics," *Journal of Political Economy* 80 (October 1972): 936–37.

essarily uninfluenced by Keynes's *General Theory*—but a strong emphasis upon the importance of changes in the stock of money for the operation of a modern economy clearly was a fundamental belief of the Chicago economists of the 1930s. Friedman did more than continue this tradition: He did much empirical work to document money's strong historical role in American economic life, and he used the theory as a powerful weapon to attack the Keynesian theory. He performed these tasks with great skill. He has an extraordinarily lucid mind. His ability to think very fast and to conduct himself with complete propriety in the heat of debate makes him an extremely formidable debater in person as well as on paper. He is a marvelous empirical worker, prepared to isolate what he believes are the essential elements of a problem, and to bring the analysis to bear most ingeniously upon empirical data. Finally he is quite talented in outraging his intellectual opponents, who have accordingly devoted much energy and knowledge to advertising his work. His only flaw as a debater, in my opinion, is that often his victories are temporary; the defeated adversary will slink off muttering "I'll think of a reply to his argument in a few days."

Milton is in no way a trickster: He believes what he says and says what he believes. When we were at the Center for Advanced Study in the Behavioral Sciences from 1957 to 1958, we staged an amusing debate (against Melvin Reder and Robert Solow, if I remember correctly) on "Resolved: That Crime Does Not Pay." I proposed a cute if fallacious economic argument for our (affirmative) side but Milton refused to allow its use. "Even when joking," said Milton to me, "an economist may not use economic analysis fallaciously." That seems pretty strict, considering how many solemn fallacies one encounters.

So more than three decades of powerful work in monetary economics were carried out at Chicago. It included major works by Friedman: a fundamental revision of the interpretation and measurement of a central Keynesian concept (*A Theory of the Consumption Function*, 1957); a brilliant analysis of the monetary history of the United States (*A Monetary History of the United States* [with Anna J. Schwartz], 1963); an influential essay on free foreign exchange markets; and extensive studies of the demand for money. Able students, foremost Phillip Cagan and David Meiselman, made important contributions to this work. The monetary side of Chicago economics was the creature of Friedman.

Friedman's work on public policy outside the monetary area is wide-ranging in both topic and media. His *Capitalism and Freedom* (1962) presents a lucid and comprehensive case for laissez-faire, and has sold over 500,000 copies. The later *Free to Choose* (1980), written with his wife Rose (sister of Aaron Director), was a much greater success both as a TV show and as a book. A vast number of lectures and debates and decades as a *Newsweek* columnist have made him a national figure in public policy. Two examples of his inventiveness as a policy analyst may be given.

Milton Friedman pioneered the proposal of the use of vouchers to give parents freedom of choice in the schooling of their children and to introduce competition into elementary and secondary public schools. The parents of each child would receive a voucher equal perhaps to the average annual cost of a student at a public school, and the voucher would be tenable at any school, public or private, of acceptable quality. This proposal strikes me as eminently desirable to improve education, but that view has not been shared by the powerful public school establishment.

The proposal of a negative income tax on low incomes is a second example. Here the proposal is to replace a vast arsenal of public welfare programs (unemployment benefits, food stamps, housing subsidies, health subsidies, etc.) by direct cash payments. The assistance to the poor could be much better targeted—now much public housing and other benefits are going to the middle classes. More importantly, the beneficiaries of the negative income tax would be given freedom of choice—to choose perhaps more food or medical care at the cost of less housing. The chief opposition to this proposal has come from the many people who believe that the poor would spend most unwisely. I, like Friedman, do not share this fear, but I do share the fear many have that the negative income tax would be added to, not substituted for, the vast smorgasbord of special policies now in place.

Both school vouchers and negative income taxes exhibit Milton's ability to present public policies in a fresh and illuminating manner. There are many knee-jerk conservatives as well as liberals, but Milton's reactions are cerebral and they are not stereotyped.

The careful and persistent use of modern price theory, the third area to which Milton also made major contributions, has a different source: Viner's famous course in economic theory was surely the origin of this part of tradition. Primarily by a rigorous exposition of price theory, Friedman instructed generations of students in its use. But let us first look at the cast of other characters of the Chicago School.

The Membership of the School

I physically joined, or was repatriated by, the Chicago School of Economics in 1958; intellectually I no doubt had long been a member. I was invited back by Allen Wallis, then dean of the Graduate School of Business. He offered me a luxuriously upholstered chair, the Charles R. Walgreen Professorship of American Institutions. Mr. Walgreen, Senior, had withdrawn his niece from the University of Chicago in 1936, and accused the University of teaching such subversive doctrines as free love (it certainly has gotten less expensive) and communism. The legislature of Illinois had created an investigating committee after the case was heated up by the *Chicago Tribune*. The University was exonerated, Mr. Walgreen was convinced of its innocence, and he gave $500,000 for the chair in American Institutions. I became the first holder of the chair some twenty years later, and it yielded me a princely salary that was the talk of the economics profession, $25,000 per year. In 1988 dollars that would be well over $100,000, still an eyebrow lifter. I joined the Department of Economics as well as the Graduate School of Business, a congenial combination because of the unusual quality and intellectual purpose of the faculties of the School and the Department of Economics.

A fascinating variety of old and new friends made up the "school." Aaron Director had succeeded Henry Simons at the Law School in 1947, and he played a major role in the Chicago School. We had become close friends at the first meeting of the Mt. Pelerin Society in 1947 and have remained so ever since. Aaron is a courteous and cultured gentleman with a sharp logical mind and a probing intelligence that thinks its way to the bottom— or at least, to an unfamiliar depth—of many questions. I gradually learned that when he began asking simple questions about some comfortable belief I had proposed or more likely simply repeated, the odds were high that I would end up with a different view of the matter.

In some ways a new friend, Reuben Kessel, was the very opposite of Aaron. Reuben was (he died, all too young, in 1975) a vivid character—direct, often painfully blunt, a bewildering mixture of naiveté and the tough strength of hard experience. He came from a family that was poor. I remember his once remarking, as we passed some delicious-looking pies in a delicatessen line, that the interval between when he could not afford them for financial reasons and when he could not afford them for caloric reasons was deplorably short. His skepticism was almost pathological. He told me once that when away on a trip he had an attack of appendicitis, and when he met the surgeon on the way to the operating room he asked what his medical credentials were. "For heaven sake, Reuben," I expostulated, "what would you know about the credentials?" "Not much, but if he got on his high horse, I would have called off the operation." Fortunately for Reuben's health the surgeon did not mount a steed. Reuben always attempted to be the true economic man, rational in all his undertakings. Of course he banked in Missouri, where he found a then rare bank that made no service charges. And, of course, he could become hopelessly emotional over matters of pride such as the rare receipt of an offer from another school. For non-economists, I must add that he was a fine economist, specializing chiefly in health economics. He caused a vast commotion by an early article arguing that hostility by the American Medical Association to Jewish doctors had been based upon their price-cutting propensity.

Ronald Coase, who came to Chicago in 1964, is the author of the extremely famous theorem which I have already discussed. Ronald is English to the tips of his fingers, a natural recluse (he has never had a listed telephone number to my knowledge), and he is a witty and elegant scholar. He is as immune as anyone I know to fashions in thinking, including economic thinking. His independence is honorably illustrated by his failure to share our profession's hopelessly exaggerated opinion of the role of monopoly in economic life. Adam Smith said in a passage, the tenor of which was not complimentary, that England was a nation of shopkeepers: I once told Ronald that with his independence and ability he was at least the proprietor of a large department store.

I could continue the inventory. There was H. Gregg Lewis, the pillar of the economics department. He was the person who solved the administrative problems of the department and the academic problems of the students, all the while reconstructing labor economics in its modern form. There is James Lorie, a pioneer in modern financial economics and the owner of the most sardonic wit I have met (although Edward Shils is a close rival).

There is Harold Demsetz, whose eventual departure for California is the only complaint I can make of a friend and colleague. There is Lester Telser, who combines great technical skills with even greater irreverence for received wisdom. There is Sam Peltzman, once a gifted student and now a superb colleague, whose scholarship is creative and as colorful as his taste in clothing. There is the brilliant lawyer (and excellent economist) Richard Posner who almost single-handedly has created the field of the economic analysis of the law. His remarkable abilities have allowed him to create an enviable career as a professor and launch another as a federal appellate judge. But these are surely enough to suggest the quality, the diversity, and the joy of a warm intellectual community. Of course there were other colleagues of distinction who worked in fields in which I was not active: Arnold Harberger in public finance; D. Gale Johnson and Theodore Schultz in agricultural economics; and all three in economic development.

George P. Shultz came to Chicago a year before I returned. He came as a professor of industrial relations but his integrity, good judgment, and administrative skills were soon widely recognized and the faculty of the Business School persuaded him to become its dean. (Someone once asked me how it could be that George accepted our deanship while declining one at a well-known eastern school. My reply was that there is no inconsistency in declining to be mayor of Boston but accepting the presidency.) We became close friends in spite of the fact that he beat me regularly at tennis and even more decisively at golf. I used to threaten to accuse him of getting a kickback on my salary because he won so many golf bets.

Milton Friedman had a substantial influence on George's views of economics. When George was Secretary of the Treasury in 1973 under President Nixon, he allowed the foreign exchange value of the dollar to float, and I believe that Milton's arguments had contributed to that eminently desirable act. Many people assign a large role to luck in their lives and I wonder if this applies to George. One could argue that if George had not served for a year in Arthur Burns's Council of Economic Advisers, Burns would not have persuaded Nixon to make George his Secretary of Labor. That surely is a gross simplification. Burns could not have been able, or even have desired, to help Shultz obtain that position, unless Shultz had established a strong reputation in the labor field and in administration. Whatever the accidents of politics and personal relationships, Shultz would have become an important figure in academia or business if he had not in politics.

I shall concentrate hereafter on the areas in which I worked, industrial organization (the economics of markets and industries) and public regulation.

The School's Work in Microeconomics

In the late 1940s there were two dominant schools of thought on the study of market behavior, both centered at Harvard University. The first was the theory of monopolistic competition, created and led by Edward Chamberlin. The second was a school of institutional market studies headed by Edward S. Mason, with Joe S. Bain, Carl Kaysen, and Donald Wallace its most active members.

The theory of monopolistic competition had wide influence in economics until the end of the 1950s by which time it became apparent that the doctrine was exhausted. That is to say, numerous economists could play variations on its central theme—that the differences *between* products of the same class (differences in quality, taste, location, associated services, etc., of, say, breakfast cereals or trucks) were prevalent in industry, but their studies yielded no interesting empirical insights on the workings of industry. The theory was descriptive, not analytical.

Edward Mason's group at Harvard was the leading producer of Ph.D.'s in industrial organization. Each new Ph.D. gravely decided in some mysterious fashion whether the industry chosen for his doctoral dissertation was or was not acting in a socially desirable way. Even the most influential work of this school (Joe Bain's *Barriers to New Competition*, 1956) was fundamentally arbitrary in the sense that it did not make essential and persistent use of modern price theory.

The emerging Chicago tradition challenged both of these ruling views. It proceeded from the assumption that modern price theory is a powerful weapon in the understanding of economic behavior, not simply a set of elegant theoretical exercises suitable for instruction and the demonstration of one's mental agility. In particular, primarily under the influence of Aaron Director, we moved away from the assumption that monopoly was almost ubiquitous in modern economies. This Chicago orientation had three main facets. The first was that the goal of efficiency is pervasive in economic life, where efficiency means producing and selling goods at the lowest possible cost (and therefore the largest possible profit). This goal is sought as vigorously by monopolists as by competitors, and monopoly power is of no value in explaining many phenomena which have efficiency explanations.

An example is the practice of resale price maintenance —the enforcement of minimum prices on retailers by the manufacturer of a product. The traditional theory had explained this practice by some such argument as that manufacturers, by eliminating price competition among retailers, also eliminated indirect pressure for price competition among manufacturers, or alternatively, that the practice was simply insisted upon by a cartel of retailers. The Chicago explanation, developed by Lester Telser, was quite different. Consider a discount house competing with a traditional department store in the sale of household appliances. Without resale price maintenance, consumers could examine the appliances in the department store and then go to a low service, low price discount house to buy them. In the long run, department stores would cease stocking a variety of appliances and providing retail services such as explanation of the properties of the products, and the sales of all appliances would fall. To eliminate this free-rider practice of discount houses, some device such as resale price maintenance was necessary.

Advertising is another example. For most economists before 1960, advertising was a device to attach buyers to one's own brand of a commodity, and was inevitably manipulatory. In particular, it was almost universally believed that there would be no advertising by fully competitive firms or industries. My work on the economics of information led to a radically different view of advertising. Consumers (and for that matter, everyone else in economic life) require much information: What new products are available, where can new and old products be purchased (since stores come and go), how can one get assurance of quality, who sells at the lowest price, and so on. Advertising proves to be an extremely efficient way of conveying much of this information, and is as essential for competitive firms as for monopolies.

A second main theme of the Chicago School of industrial organization is that it is virtually impossible to eliminate competition from economic life. If a firm buys up all of its rivals, new rivals will appear. If a firm secures a lucrative patent on some desired good, large investments will be made by rivals to find alternative products or processes to share the profits of the first firm. If the state gives away monopoly privileges (such as TV channels), there will emerge a strong competition in the political area for these plums.

Harold Demsetz made an important contribution to this argument by analyzing the so-called natural monopolies: public utilities such as water companies, which are given exclusive rights to supply a city. He pointed out that here the force of competition would be felt at the stage of granting the franchise; there would be many bidders for the right to supply the water and their competition could be directed to the consumers' benefit.

The strength of the competitive force does not imply that there are no monopolies or that they are always transitory, although in century-long periods they are indeed transitory. What is implied is that the processes of obtaining, defending, sharing, and eliminating monopoly positions are more important and interesting than the exercise of monopoly power. John R. Hicks once said that "a quiet life is the best of monopoly profits," which reveals more about the life of a great Oxford economist than it does about the marketplace.

The combination of the shift in attention to efficiency, and the restoration of the powerful role of competition, has done much to weaken the arguments for an antitrust policy that seeks to deal with minor or transitory or (as in the case of vertical integration) erroneously identified monopolistic practices. A major antitrust action usually requires five or ten years from start to finish and costs plaintiffs and defendants many millions of dollars. (It was said in the IBM case that the budget by the company for the defense was not limited, but nevertheless exceeded that limit.) Antitrust remedies should be reserved for important and persistent monopoly problems, many or most of which are created by governmental regulations.

A third facet of the Chicago School's work was on the theory of public regulation. Here, oddly enough, the work departed from the ruling tradition in an unusual way. The vast literature on public regulation of industries, products, and prices, was overwhelmingly devoted to how it *should* be done (called normative analysis). We initiated the study of *why* acts of regulation were undertaken and what their effects actually were (called positive analysis). I have discussed some of my studies in this area in chapter 7.

All three of these Chicago approaches—the shift to studies of market efficiency, the pervasiveness of competitive pressures, and the application of economics to the causes and effects of public regulation—have had strong influence upon the literature of industrial organization.

Indeed it can be said that they conquered the field; by 1980 there remained scarcely a trace of the two Harvard traditions of Chamberlin and Mason in the current work of economists.

Recent Chicago Economics

In the last fifteen years new leaders of Chicago economics have emerged and given the school new directions of research. The foremost two of these economists have been Gary Becker and Robert Lucas, Jr., but there have been a number of other important additions such as Sherwin Rosen and Edward Lazear.

Becker is a most creative economist who has immensely widened the domain of economics. After a doctoral dissertation, the first to apply economic analysis to racial and sexual and other forms of labor market discrimination, he became the leading figure in the development of the concept of human capital (the value of the talents and skills of a person, and the forms of investment by which they are created). Thereafter he revived the economic theory of crime and punishment and created the economic theory of the family (marriage and divorce, number and quality of children, altruism, etc.).

Lucas was a pioneer in the development of the theory of rational expectations in macroeconomics. The essence of this approach is the appreciation that the actions of governments (and equally of private persons) do not come as surprises to economic actors. If the Federal Reserve commonly engages in the sale of bonds to banks when a strong inflation seems imminent, then the entire financial community learns to expect this action, and to take appropriate actions to insulate itself from the effects of the restrictive monetary policy even before the policy is instituted. This theory has raised havoc with much traditional macroeconomic theory including Keynesian theory.

Do these new developments represent a recognizable continuation of the main elements that constituted the core of the Chicago School? I believe that they do. Each is a persistent and consistent application of fundamental economic theory to a range of institutions and behavior previously considered to be "given" to the economist: opaque facts of life instead of the products of rational economic behavior. To put it differently, it would have been surprising to see the appearance of the works of two economists such as Becker and Lucas and their followers at a university hostile to the traditional Chicago School.

The Future

A school of thought must have a limited, and often short, life. It must either convince the discipline to accept its central propositions, in which case its reason for existence vanishes, or it fails to convince, and then futility and boredom end its days. The Austrian School persisted from 1870 until 1930, a remarkably long span, because its rival, the German Historical School, continued to preach an antitheoretical methodology for economics. The German School, in turn, persisted despite its lack of permanent influence on economics because it controlled the Ministry of Education, which in turn controlled all professorial appointments.

The survival of the Chicago School for forty-odd years is a tribute not to the obstinate tenacity of the school but to its inventiveness. Friedman's departure in the mid-1970s was compensated by the new lines of work of Becker, Lucas, and the public regulation group.

Even when a school wins over its rivals, as Chicago did in the area of industrial organization from 1960 to 1980, the victory is not permanent. Nor should it be if victory involves freezing research in a new pattern. The most recent development in industrial organization is the emergence of game theory, which has dominated the writings of the younger economists in the major eastern schools and Stanford. This literature is closely related in spirit to Chamberlinian economics: It is much more rigorous (as well it should be, fifty years later) but has not shown equal gains in empirical motivation or empirical applicability.

Once a school of thought has been well-established, the recruitment of new faculty tends to reinforce its central tendencies. This may be due simply to the practice of recruiting scholars sympathetic to the school's views, but another force is even stronger. In the last two academic generations, Chicago has offered positions to Paul Samuelson, Robert Solow, and James Tobin, a group slightly younger than my generation, and to a still younger group, Robert Barro, Stanley Fischer, Robert Hall, Dale Jorgenson, and Thomas Sargent. Most of the members of this distinguished list, I conjecture, declined appointment largely because they felt that here they would be immersed in an alien intellectual environment. They may possibly have been wrong (a little earlier an avowed socialist such as Oskar Lange seemed to do well in our environment), but one cannot be certain.

Schools emerge in response to scientific need; they are not created in some social compact. That must mean that they serve an important scientific function—to bring together a group of scholars who share a common view of the proper new direction of their science. The group strengthens their common view of the science's proper agenda by self-criticism, varied applications, constant emendation, and usually vigorous controversy with rival research agenda ("paradigms"). A major reorientation of a full-scale discipline almost invariably calls for the efforts of several scholars: Even an Isaac Newton or an Adam Smith required disciples and fellow workers to conquer an established doctrine or methodology. I consider, with no pretense of objectivity, that the Chicago School has been an important source of the prosperity of American economics.

CHAPTER 11

Academic Freedom
and Responsibility

IT WAS almost exactly five years after I had left Iowa State College that it became the scene of a hot controversy over the rights of academic scholars. Oswald Brownlee, a superior young economist who was still working for his Ph.D., wrote a pamphlet in the College's Wartime Farms and Food Policy Series. The pamphlet was called *Putting Dairying on a War Footing*, and ironically enough, it succeeded in doing so immediately so far as Iowa dairy farmers were concerned.

Brownlee proposed a method for dealing with the war-induced decline in the output of dairy products. The central theses were simple:

Even though some of the milk solids are lost in the whey, cheese is a concentrated and economical food.

Butter is in a somewhat different class. It is a high cost fat; and only a small part of the skimmed milk, a by-product of its production, goes to human food. Vegetable as well as some other animal fats can be produced at less cost of manpower and other resources. These can be used in margarine to make up the butter shortage. . . .

Re-examine the allotment of fats and the allocation of materials for manufacturing facilities for margarine so that consumers will have a substitute for butter. Restrictions on the sale of margarine—state excise taxes, license fees, etc.—should be removed so that its consumption may be encouraged.[1]

These arguments were developed and supported, and to an economist they appeared compelling.

That the Iowa State faculty and administration were surprised by the violence of the reaction of the dairy interests suggests a good measure of naiveté. The dairy industry had already established a record of ruthless opposition to the use of oleomargarine. Since 1902 the federal government had put a heavy tax on colored margarine (its natural color when made from beef fat was white). That tax was not removed until 1950. Some states prohibited the sale of colored oleomargarine, and some required restaurants that offered margarine to put up signs with letters at least two inches high stating "Oleo-

[1] Oswald H. Brownlee, "Putting Dairying on a War Footing," original (retracted) version of Pamphlet No. 5, *Wartime Farm and Food Policy Series* (Ames, IO: Iowa State College Press, 1943), 1–2, 3; quoted in Muriel A. Weir, "Pamphlet No. 5 and the freedom to publish at Iowa State College" (Master's thesis, Iowa State College, 1976), 204, 206–7.

margarine used here" or "Imitation butter used." Iowa was the second largest butter-producing state in 1943, and was not prepared to let the College forget it.

The controversy is carefully recounted in an unpublished 1976 master's thesis by Muriel A. Weir, *Pamphlet No. 5 and the freedom to publish at Iowa State College*, and I do not intend to repeat its details. A weak college president forced the withdrawal of the pamphlet and a year later a revised edition appeared. That revised edition was suitably more circumspect in its language, but thanks to the integrity of the author and the support of an able dean, Robert E. Buchanan, none of the essential points in the first edition was withdrawn. Professor T. W. Schultz, head of the Department of Economics and Sociology when the pamphlet was recalled, resigned and went to the University of Chicago.

Of course I shared the outrage of the economists that we should be limited in our freedom to publish works of scholarly quality—and, I surely assumed, *we* were the people who should judge this quality. I should have recognized one small element of chutzpa in the proceedings: that the college press, supported in part by the dairy industry, was publishing the pamphlet.

Brownlee's proposal was a sound proposal, and it has been broadly adopted by the American nation through the gradual withdrawal of most legal obstacles to the sale of margarine. The dairy industry still receives many legislative favors, of course, and I once recommended that a new award, The Golden Udder, be given to Senator Proxmire for his indefatigable services to that industry.

I recount an instance of censorship I personally experienced. David M. Blank, then a recent Columbia Ph.D., and I wrote a monograph, *The Supply and Demand for Scientific Personnel*, for the National Bureau of Economic Research. It was written in 1956, and published in 1957, the year in which there was a wave of fright in America over the successful Soviet launching of Sputnik. The monograph went through the usual careful review at the National Bureau and was sent to the Princeton University Press, which then published the Bureau's books.

The Princeton faculty reading committee contained an engineer, and presumably under his influence it demanded the removal of a half dozen pages from the manuscript. The offending pages argued that there was no apparent shortage of scientific personnel in the United States, judging from the fact that their earnings had not risen—and indeed, had fallen—relative to earnings in other professions. Both the National Bureau and the Princeton Press were unyielding and the book was published elsewhere. Since we were employing the economist's basic measure of scarcity, I believe now, as I did then, that we were correct. The objections of the Princeton reading committee were surely to the potential relevance of our argument to public policy, though we did not argue that federal aid to train engineers and scientists should be reduced. Of course, the episode really has no lessons beyond the two—that professors of engineering (or any other subject) are just as eager as dairy farmers to protect their own particular interests and that university presses have their moments of weakness.

Professors believe that they should be free to express their opinions; not only free of penalties on themselves for expressing an unpopular opinion but also free of penalties on their institutions. That is asking quite a lot: It is asking people who have the power to punish a university by withholding contributions or by other actions to accept attacks upon themselves without retaliation.

The argument for this professorial immunity is the argument for free speech and free inquiry. If we *know* a thing is true—a released stone will fall toward the earth —there will be no harm in allowing a madman to challenge it. But most things are known less certainly, and then the person who questions prevailing beliefs may help us to correct them or to understand more fully their invulnerability to the criticism. In the words of Edwin Cannan, an English economist, "However lucky Error may be for a time, Truth keeps the bank and wins in the long run."

If truth wins on average, the preservation of the university as a sanctuary is a wise social policy; it emboldens scholars to enter into risky thoughts and schemes. Organizations that need to win every time, not simply on average, are extraordinarily vulnerable to error, and they seek to protect themselves by avoiding all risks of serious error. Governmental bureaus often face this problem: A single catastrophic error (granting, say, permission to sell a drug that proves lethal) will cause heads to roll. As a consequence, they choose a policy that will avoid large risks—and large gains.

It is only reasonable that much more freedom to err is given to the more conspicuously successful institutions. Harvard University can survive a much bigger display of error or nonsense by its professors or students than can an ordinary school, and has exercised the right. The rule applies also to individuals: A major scholar is apparently entitled—not to say, encouraged—to take strong positions on subjects far removed from his professional expertise.

The likelihood of error and the magnitude of the error will both be larger, the more careless (ambitious?) a scholar becomes. Serious error could almost always be avoided by having work carefully reviewed by fellow professionals. (Of course, serious originality would also encounter trouble in the process.) Frank Knight once proposed that a professor should be allowed to publish a piece for public consumption only after it has met the approval of his colleagues at the university. He argued that a professor should not be allowed to influence the lay public if he could not get his colleagues' approval of the professional quality of his ideas. Knight's proposal was viewed indulgently as a cute piece of creativity, and I have no doubt that it was pointed out at the time that this particular proposal had not been approved by his colleagues in the Department of Economics at the University of Chicago.

But Knight had a point. Professional articles are required to go through something like the proposed review before publication in the major journals in a field. My journal, the *Journal of Political Economy*, sends a submitted article out to one or two experts (referees) for detailed appraisal. Then the editor makes up his mind on the article's promise—the author is almost invariably asked to revise it if it is promising—and acceptance of the article requires the agreement of a second editor. The process leads to the rejection of at least nine out of every ten articles submitted, although of course some of these articles will eventually appear in some other (in our mind, lesser) journal.

However, if the economist writes for the *New York Times*, the *Wall Street Journal*, *Harper's* or the *Atlantic*, or *Penthouse* or *Playboy*, the quality standards are not imposed by fellow professionals. Instead the standards of professional journalism are imposed: They include better writing and a deep, insatiable infatuation with controversy.

Academicians love to condemn the popular press—that is, the press with many readers. It is certainly true that the economics that appears in *The New Yorker* is often bizarre, and that which appears in the extremist press is often outrageous. But these journals are conducted to please their readers, and almost invariably the readers of each popular periodical have a distinctive set of interests and preferences. The *New York Times* was shocked by the idea that the city's municipal extravagance should be punished by bankruptcy; and, indeed, shocked also by the perfidy of the American people in electing Ronald Reagan. When professors enter this arena, they must dress and fight as the local gladiators do.

I do not know why academicians require or deserve relative immunity when they leave their area of professional competence to write or talk on other subjects. If the boundaries of competence could be drawn clearly, there would be a good deal to say for letting professional entrepreneurs take their profits and losses like other entrepreneurs when they went outside these boundaries. Probably the best way to approach this state would be to require academic scholars to publish their nonprofessional work anonymously. That practice would serve two ends. Anonymity would deprive the work of an authority that is attached to the author's professional status and probably should not be attached to writings outside his area of professional competence. Moreover, the employing institution would not be drawn into the maelstrom of current events. This recommendation may sound utopian, but it was how the premier journals of Great Britain were conducted in the first half of the nineteenth century. The greatest of those journals, *The Edinburgh Review*, preserved anonymity from 1802 to 1912.

No sharp line, however, can be drawn between one's professional competence and the rest of one's knowledge and opinion. An economist can criticize protectionist policies in general with the authority of two centuries of development of the theory of international trade. He will have much less professional knowledge to guide him if the problem is to devise policies to remove protectionist farm policies of Japan or the European Common Market, and he will be almost helpless in telling the Congress how to resist domestic protectionist pressures.

The proposal of anonymity in public writing assumes that scholars will often be irresponsible in such writing, and seeks to make the best of that situation. An alternative would be to instill a code of self-restraint that would lead scholars to deal responsibly with current issues, even when those issues are of critical concern and enlist powerful emotions and interests. I find that prospect even less practicable than anonymity. Not that anonymity is such an easy route. Londoners read in the January 1830 *Edinburgh Review* the following opening of a review of Poet Laureate Robert Southey's *Sir Thomas More*:

> It would be scarcely possible for a man of Mr. Southey's talents and acquirements to write two volumes so large as those before us, which should be

wholly destitute of information and amusement. Yet we do not remember to have read with so little satisfaction any equal quantity of matter written by any man of real abilities. We have, for some time past, observed with great regret the strange infatuation which leads the Poet Laureate to abandon those departments of literature in which he might excel, and to lecture the public on sciences of which he has still the very alphabet to learn. He has now, we think, done his worst. The subject which he has at last undertaken to treat is one which demands all the highest intellectual and moral qualities of a philosophical statesman, an understanding at once comprehensive and acute, a heart at once upright and charitable. Mr. Southey brings to the task two faculties which were never, we believe, vouchsafed in measure so copious to any human being, the faculty of believing without a reason, and the faculty of hating without a provocation.[2]

They surely said, "Tom Macaulay is at it again!" But journals shouldered the blame and prospered on the credit for these writings.

I reluctantly conclude that nothing will change. Individual scholars will continue to display every degree of responsibility and irresponsibility in their writings for the public. (Those Nobel laureates who issue stern ultimata to the public on almost a monthly basis, and sometimes on no other basis, set a particularly regrettable model.) There will be acts of retaliation against them if they are in weak

[2] Thomas B. Macaulay, review of *Sir Thomas More; or Colloquies on the Progress and Prospects of Society* by Robert Southey, *Edinburgh Review* 50 (Jan. 1830): 528.

schools and against their institutions if they are in strong schools.

If we could decrease our entanglement in contemporary policy issues, whether by anonymity or self-discipline, we would surely increase the authority of economics. Then economists' judgments, say, on the efficiency of a given policy in achieving its goal, would not invite the often correct suspicion that professional knowledge was being used for partisan purposes. The tendency of scholars, alas, is in the opposite direction. The once aloof field of physics is now crowded with reckless spokesmen on every issue of the day.

I shall be so bold as to assert that it is more important that a society be intelligent in its economic policies than in its use of nuclear power. In Russia alone, millions of peasants were killed by the economic policies of the 1930s—many multiples of the effects of the Chernobyl disaster. I hope for our society's sake, therefore, that we will become increasingly more professional in the use of our knowledge.

CHAPTER 12

Academic Life on the Battlefront

A UNIVERSITY COMMUNITY has been an enclave in modern society, a group of teachers who live in important respects like members of a medieval monastery. Surely not in vows of poverty and chastity, but nevertheless in a continuously intimate relationship in which nonacademicians play relatively unimportant and transitory roles.

The situation has been changing in the great urban universities (and most great universities are in or near large metropolitan communities), as more and more of the professors are engaged in consultation and other outside activities, so that on any given week day possibly one-tenth of them are away from the university. Fifty years ago that kind of external activity was most unusual among the professors; they spent their days and their nights in company with one another.

When professors' whole lives were spent in residence, so to speak, their important social as well as intellectual associations were university bound. Each professor knew most of his (and rarely, her) senior colleagues, whatever their fields of study, and certainly he knew well the junior colleagues within his department of study. Living in such constant intimacy, strong friendships and sometimes strong hostilities emerged. I wish to report an episode in one of those engagements.

The two parties to the quarrel were both eminent economists at the University of Chicago, and although I have already introduced them, they deserve a fuller introduction. One party was my teacher, Frank H. Knight. This son of a large, poor Illinois farm family was a fascinating character. His fierce independence is illustrated by an episode recounted by one of his younger brothers. Under the urging of their deeply religious parents, on one Sunday the numerous children signed pledges to attend church the rest of their lives. On returning home, Frank, then fourteen or fifteen, assembled the younger children behind the barn, built a small fire, and said, "Burn these things because pledges and promises made under duress are not binding."

After a strange educational career dictated by the family's poverty, Knight received his Ph.D. from Cornell University in 1916 at the age of thirty, and, after almost a decade at the University of Iowa, came to the University of Chicago in 1927. By then he was already famous for a brilliant book on economic theory (*Risk, Uncertainty and Profit*, 1921), and also for a set of highly provocative philosophical essays (reprinted in *The Ethics of Competition*, 1935). He was, and remained, a strongly individual thinker, skeptical of much economic theory and of the integrity and quality of all public discussion of economic and political problems.

The second party was Paul H. Douglas. Douglas and Knight had some parallels. Douglas's youth also was spent in poverty and, indeed, in a sadly broken home. Both men were, in different ways, great economists. Where Knight was the subtle theorist, seeking to reach deeper levels of economic analysis and social philosophy, Douglas was an imaginative empirical worker and a buoyant, hyperactive liberal reformer. If Knight had written an autobiography, it would have been a tale of scholarly academic life, and little else. Douglas did write an autobiography, *In the Fullness of Time* (1972), which devotes a few perfunctory pages to his academic career and more pages to his efforts to bring Samuel Insull to justice. For him the climax of life was most certainly eighteen years in the United States Senate. After he involuntarily left the Senate I invited him to give a lecture at the University. When people told him how well and rested he appeared, he responded that he would rather be a tired senator.

The expansive, wide-ranging liberal and the intellectual iconoclast had little in common, and only the compulsory proximity of membership in a small department (about ten faculty) would lead to their association. After Knight's arrival in Chicago in 1927, his contacts with Douglas must have been fragmentary at first. For example, Douglas took six months' leave at Swarthmore College to study unemployment in 1930, and soon was on a variety of public commissions. Yet the frictions with Knight were building up, and the two men eventually resorted primarily to letters in communicating with each other (hence my knowledge). One must assume that the following letter was not the opening salvo:

December 21, 1933

Professor Paul Douglas
Faculty Exchange

Dear Paul:

The remark you made the other day on the way over from the Club, about pleasing your colleagues by shortening your life, didn't quite sink in until you were out of sight. Since then it has rather "stuck in my craw." I have felt for some time that you and I seem to be developing a tendency for each to feel a little "off" at the other, chiefly, it seems, because each feels that the other feels that way. At least I don't know any other reason. We disagree pretty widely, I guess, about the role of scientific economics and economists in the current political situation. Frankly, I think that those who go into politics, particularly under present conditions, are committing suicide on behalf of the profession as a whole. But on the other hand, I don't pretend to have anything like knowledge on the subject, and on the other I don't see that any such difference need be the occasion for any lack of personally cordial feeling. For instance, T. V. Smith is at least as much of a barnstormer and uplifter as you are, and he and I talk about the issues with complete amiability. As a matter of fact, I have rather tried to cherish an opinion of humanity, at its best at any rate, according to which men might feel compelled to fight to the death over an issue, and still keep the most cordial personal feeling and attitude. How about having lunch, or something of the sort, and talking about it some time?

Cordially,
Frank H. Knight

The response was quick:

December 22, 1933

Dear Frank:

I have not been aware that I am lacking in cordial feeling toward you. It is always dangerous to write about such matters, but it has been quite evident from your personal conversations and your public speeches and what you have said to many others that you regard me as something of a charlatan and a demagogue. I am not wholly stupid, although probably partially so, and therefore what you actually think of me has not failed to sink in. Under the circumstances therefore it seemed to me a quite proper comment that it would be a great relief to some of my colleagues if by working on the Consumers' Council I did shorten my life.

But I can say in response that I do not want you to shorten yours, since I have the highest opinion of your intellectual ability and the great contributions which you can make to economic science. In short, my attitude has been that of one, who not knowing why he should be let in for a rather severe public and private denunciation under the circumstances has decided that the best thing to do is to keep his mouth shut and refrain from recriminations at all costs.

If you really think a personal meeting would clear up any of these matters, I should of course be very glad to meet with you in a completely sincere and friendly spirit, but I do not wish to exacerbate the situation. I too have hoped that we might have cordial and friendly feelings toward each other, but it seemed to become so evident that you did not hold such an attitude towards me that I felt I was irritating you by my existence, and much more by my presence.

In conclusion may I in no ironic spirit and with all sincerity wish you a merry Christmas?

Faithfully yours,
Paul H. Douglas

Knight replied:

Dec. 27, 1933

Dear Paul:

The tone of your letter is quite a shock. I had suspected nothing beyond a tendency to rather go separate ways, until you made the remark I mentioned. As far as I know there is absolutely no ground for bringing words like charlatan and demagogue into the discussion and I wonder what could have put them into your mind. I am deeply, even bitterly disappointed and desgusted [sic] with the status of economics and intellectual leadership and even more with conditions within the political "science" professions—but have never blamed any individual—I have had my part—in proportion to my weight no doubt—in what has happened. I think it has been very unwise, but I doubt whether we could have had it different really, and *I* think we are sunk anyway. I do not contend that it is either illogical or wrong for the individual to try to influence the course of events, if he thinks he can. I only *wish* that we could have a group life holding aloof from political struggle until agreement was reached on essentials—partly no doubt because in the absence of such a group I have no contribution to make, really nothing to do to earn my living. I did not take up this work with the idea that it meant a sheltered and parasitic life.

To my knowledge I have not said anything discredible of you, as to intelligence or sincerity, in public or private.

Thanks for the Xmas greetings, and may I return them for the New Year?

Very Sincerely,
Frank Knight

I was a newly arrived student in 1933 and none of this controversy leaked out to us. Of course we knew that their relationship was cool, and there was little overlap in the students who studied with each. I would find it hard to believe Knight's disclaimer of criticism of Douglas, because Knight had no capacity for silence. (In attending a colleague's course, he would say, "Don't be a damn fool, Henry.") He also believed that for a scholar to give explicit solutions to hard social problems on the basis of inadequate knowledge was almost immoral. Hence his suggestion, to which I have referred, that no professor could address proposals to the public unless that professor's colleagues approved of the proposals—a suggestion that must end in invariable pro forma approval or mass resignations.

In the autumn of 1934 the difficulties increased by an order of magnitude. Henry C. Simons, whom I have already introduced, was facing the tenure decision. Simons was Knight's disciple—he had followed Knight to Chicago from the University of Iowa—although he had a thoroughly independent and brilliant mind. He became the subject of contention.

Douglas was opposed to Simons's promotion and even his continued appointment because—the brilliant *A Positive Program for Laissez Faire* aside—he had published nothing and was not a popular teacher. In the next round of letters, Douglas became specific:

> Moreover, from time to time, a number of students have complained to me, many of them of real ability, that they find his courses unattractive and that he is almost openly insolent to those who ask questions and generally tells them that the questions they are asking are either too elementary for him to pay attention to or that they are not questions which any sensible person should propound or to which any sensible person reply. If there is a better way of killing off intellectual interest upon the part of the students, I do not know of it. I have, therefore, come to the conclusion that on the whole Simons has been a comparative failure as a teacher.

Inevitably personal relations between Knight and Douglas entered, and the latter aired some grievances:

> What is abundantly clear, however, is that for at least two years and probably more you have been attacking my motives before groups of students, classes, fellow members of the faculty and upon occasion the general public. I have abundant witnesses that in a large class where you gave lectures before one of the survey courses that you referred to me by name as a publicity seeker and contrasted your own attitude as being directly the opposite of this. In a seminar you referred to me as being a "godsaker," with the clear implication which was perhaps explicitly stated that I was a loose-witted one as well. Numerous members of the faculty have reported to me personal conversations which they have had with you and where before relatively large luncheon groups you have labelled me again and again as a publicity seeker craving to get his name in the newspapers and seeking every opportunity to do so and that the reasons why I have advocated certain proposals, which to me are dear, has been simply a desire to be talked about.

Knight replied with a denial that he had attacked Douglas: "I *know* that I have *never* done any such thing." He defended Simons in highly personal terms:

> Speaking of feelings, I feel that I must go on to what is undoubtedly a confession of weakness. I do not think that I can put this issue in cold terms of abstract advantage and disadvantage. I "feel" as if eliminating

[Simons] is eliminating me, and that when it is done I would be simply "through" with the group, morally and sentimentally. But with *absolutely no enmity* toward any individual; it would be just the nature of things. Naturally this feeling is related to the awareness that I am myself the superfluity in the situation. It is in my special field that we are topheavy anyway, and I represent nothing in particular that is not adequately provided for otherwise. Moreover, in "qualitative" terms if these men do not belong in the group, I do not belong.

The controversy was resolved, Solomon-wise, by retaining Simons and releasing another Knight disciple. We are entitled to believe that thereafter the argument subsided. By 1939 Douglas had become an alderman in Chicago, and his life moved out of academia, soon to the Marine Corps (though he was a Quaker) and then to the United States Senate.

Frank Knight was my thesis chairman, and he befriended me all his life. Yet, as I read these letters I deplore the degree to which he associated his own position and dignity with that of Simons: That is a tactic which if often used would make departmental decision making utterly impossible. Of course, professors still push "their" candidates hard, but manners are important, even essential, to civilized discourse.

I agree with Douglas that Simons's credentials did not fit the former half of a "publish or perish" policy in 1935. Simons soon did begin to write, and produced major articles on monetary policy, trade unionism, antitrust policy, and an important book on personal income taxation in the remaining decade of his all-too-short life. In addition, he initiated what has become a major development in legal education when he began teaching economics in the University of Chicago Law School. Eventually he received tenure, first in the Law School, and then also in the Department of Economics. It is nice to think that Knight may have seen Simons's promise more clearly than Douglas.

Most people find close and continuous personal argument extremely debilitating, and, perhaps unhappy marriages apart, most people find tolerable ways to escape such situations—changing jobs or places to live, for example. Douglas and Knight had been thrown by their academic positions into close and latently even competitive relationship; each eventually became president of the American Economic Association, for example, and the two had unharmonious plans for the future of the Department of Economics and, indeed, for the world.

All vigorous scholars are reformers whether they follow Douglas into the outer world or remain in academia with Knight. Knight devoted great energies to his attempt to reform his fellow economists in their views (for example, on capital theory) and equally great energies to opposing the then popular proposals for central economic planning. In fact his open disapproval of Douglas's political activities was itself the act of a would-be reformer. No one finds it easy to keep his hands, or tongue, off other people.

CHAPTER 13

The Imperial Science

ECONOMISTS have a proprietary interest in subjects such as the gross national product, unemployment, and the financial markets. They assume ownership over problems such as whether a minority's wages are being reduced by discrimination or whether a protective tariff reduces a nation's income. Of course that doesn't mean that they know all the answers, any more than medical science can answer all questions about illness. But it does mean that economists believe that their answers to traditional economic questions are at least as good as, and probably a lot better than, the answers given by any other group in society. On pain of ostracism, I of course accept this position.

One would think that was work enough for any profession, no matter how smart or diligent its members. And that is true. Economics abounds with unsettled problems, and such is the nature of research that as we make progress with a problem, the problem often proves to be deeper and more complex than we expected. As an example, in recent times a good deal of fine work has been done on the capital markets and how they accumulate capital and allocate it to various users. Nevertheless, we have no sensible answer for so simple a question as why most societies pass (usury) laws setting maximum interest rates; assuredly the laws do not help poor borrowers. Indeed, the laws injure the poor by preventing them from offering higher interest rates to compensate for the greater riskiness of their repayment of a loan, and thus deprive them of access to legitimate capital markets.

I have already commented in chapter 7 about the extensive work now being done by economists on political institutions and practices that exert important effects on traditional economic affairs. That extension of economists' work is easily explained: The state is playing a growing role, often a decisive role, in the workings of the enterprise system. The extension has good linguistic antecedents: After all, for more than a century (from 1760 to 1870) economics was usually called *political economy*. To me the real puzzle is why economists were so slow to use their theory and research methods to analyze the economic role of the state.

It will be a recurrent theme that the extension of economic analysis to other fields has encountered resistance from economists as well as from the scholars in these other fields. Critics delight in posing problems that confront these extensions; thus so unlikely a critic as Reuben Kessel once challenged me: "If political behavior is rational, why do legislatures lower tax rates before elections and raise them after elections? Doesn't that attribute ignorance to the voters?'' Kessel required at most five minutes to think up this question, and it took Claire Friedland and me two days to show that its factual assumption is mistaken: Tax rates are raised on average as often before as after elections.

That episode led me to reflect upon the nature of scientific criticism. Nothing is easier than to suggest new work to a scholar: Was the result under discussion not contradicted by French (or better, Russian) experience? Would the results change if account were taken of the ethnic composition of the population? Ad infinitum. Each such suggestion-criticism can be produced almost costlessly but may require months of time to investigate. I was led to propose that critics should be rewarded if their suggestions are fruitful and taxed for part of the investigation when they are wrong. I was told by my colleagues that my suggestion, if adopted, would reduce the scholarly community to silence. I suppose that means that the price for successful suggestions was too low.

But let me return to my theme: The very nature of economic logic invites a sweepingly wider application of economic analysis to social phenomena. An economic problem is a problem of choosing efficiently among alternative ways to use resources, whether the resources are dollars, a bowl of whipped cream, available time, or even a reputation for honesty and skill. Consider some ancient applications of this logic to problems far removed from familiar economic prices and markets.

A famous economist, Philip Wicksteed, remarked that he loved fresh eggs, so the farther out he moved from the center of the London of 1910 in which he lived, the easier it would be to keep chickens and have fresh eggs. But the farther out he moved, the fewer times his friends would come to visit him, and he also prized his friends. So, he reported, he moved out just far enough that to go a little farther would have cost him more in friends' visits than it gained him in fresh eggs. That decision utilizes a routine economic rule: Divide a resource (here, basically space) so that at the margin it yields equal returns in both uses. If we can compare fresh eggs and friendships, what can we not compare?

Such whimsical examples can be found throughout the history of economics. Adam Smith remarked in 1776 that one could infer from the fact that the Pennsylvania Quakers had emancipated their slaves, that they had few slaves. He was stating that the action was taken because it was not costly—even philanthropy reflects the law of demand: People will always buy more of anything when its price is reduced. A few decades later Patrick Colquhoun reported that agricultural lands near London were worth less than those farther out. That is exactly the opposite of what one would expect because produce from more distant farms had to bear higher transportation costs when it was brought to the city market. His explanation was that the closer a farm was to London, the more of its crop was stolen! Nowadays it is standard procedure to measure the cost of crime to an area by its effects on property values.

Such examples prove that economists have often looked outside their traditional field, but is looking outside useful? To take Wicksteed's example, of what use is his advice to equate the marginal utility of fresh eggs to the marginal utility of friendships? Surely none of us is facing precisely the same choice, and if we did, the rule would not tell us whether to locate near or far from central London or Los Angeles (if it has a center). His example was intended only to teach his readers that economic logic is the logic of all efficient behavior.

The answer, of course, if there is an answer, to this skepticism must be that these invasions by economists of other fields have yielded important understanding of social problems, and therefore have helped us to deal with them. That is the case, as I hope to illustrate. But first let us look at some of the modern imperialism of economists.

Gary Becker is *the* leader of the extensions of economic analysis into nontraditional fields. He began early: his 1957 doctor's thesis, *The Economics of Discrimination*, was the first application of economic theory to the troublesome area of racial and other forms of discrimination. He must have gotten an early taste of the unwillingness of most people to follow his lucid, inquiring mind into new regions because it was a struggle to persuade the University of Chicago Press to publish the monograph. (My contribution to the book, as I recall, was to write to the Press urging its publication.) It has become a classic in the literature on the subject.

What can an economist say about such a subject? He can, of course, approximately *measure* the extent of the discrimination in, say, wage rates, which in itself proves to be a complex task if one tries to take proper account of the quantities and qualities of the labor market skills of members of minorities and majorities. Economists have done very well for themselves as experts in the flood of affirmative action litigation in recent decades. In one wonderful study at our Center for the Study of the Economy and the State at the University of Chicago, George

Borjas found that the Department of Health, Education, and Welfare (as it was then known) was flagrantly violating its own standards for nondiscrimination. Secretary Califano admitted the violation, said it had been unavoidable, but nevertheless would soon be corrected (see the *Wall Street Journal* for July 12, 1978).

Gary Becker employed the theory of international trade in studying discrimination to reach conclusions such as these: (1) a small minority loses heavily by discrimination; a large majority loses little; and (2) retaliatory discrimination by the minority against the majority *increases* the losses of the minority.

It requires courage to deal objectively with issues as loaded with emotional commitments as racial discrimination. One is certain to be accused by some readers (and even more by many nonreaders) of moral insensitivity if not downright prejudice. Yet the social value of Becker's scarce combination of courage and great analytical power is immense; only by studies such as his are we able to understand the relationship between the goals we seek and the public policies we may undertake. A striking example of the recent need for such work is the issue of divestment by colleges of their investments in companies dealing in South Africa. There are at least 1,000 pages of appeal to racial justice for each paragraph of analysis as to whether divestment would affect the amount of foreign investment in South Africa, and, if it did, whether the condition of blacks in South Africa would be improved or injured.

Soon Becker turned to the decisions of families as to how many children to have, and offered the postulate: "Children are viewed as a durable good, primarily a consumer's durable, which yields income, primarily psychic income [pleasure], to parents." Armed with a theory of consumer demand developed to deal with refrigerators and automobiles, Becker was able to analyze problems such as the choice between number versus quality of children (quality being the amount spent on each child for health, education, and other items). In particular, he predicted that once people possessed reliable knowledge of contraception, the number of children would increase with family income—a reversal of the age-old pattern.

I attended the conference at which Becker's paper was presented in 1960. I still remember the tone of outrage with which a Harvard economist complained at the impropriety of comparing babies with refrigerators and other durable consumer goods. Yet the economic analysis of fertility has become a vigorous and important area of work, and it has deepened and widened the understanding of population changes.

If childbearing seems an unsuitable subject for economists, the subject of marriage must appear doubly so. The importance of the subject, however, justifies the risk of outrage. Even in this age of high and rising divorce rates and extensive nonmarital cohabitation, marriage is the most durable and important personal relationship between people. The marriage contract—it is a contract by law even if the couple enter no other written agreements —is fundamental in every society; it provides the institutional framework for the perpetuation of the human race.

It should not be surprising that Becker based his treatment of the subject on the assumption that males and females choose their mates with a view to maximizing their income: No other available match promises so much to each. Becker defines "income" broadly, of course. It comprises an entire life-style including children, a comfortable home, and a desirable social life. In the marriages of a hundred years ago, the wife's contribution was usually within the household, but often included production for the market on farms and in small businesses. Nowadays the majority of adult women are in the market labor force, and their households have larger money incomes and less of the other components of "full" income. So income is more than dollars. It is probably needless to report the antagonism this approach encountered. The critics ask: How can one treat in so calculating a fashion the most intimate of human relationships and the love that gives rise to it? I must confess to the opposite reaction: Is it not proper that so deep and lasting an association should be reached through rational thought? Does this forethought not add dignity to marriage? Would one expect much of a partnership that was formed without regard for the long-run consequences of that association?

Whatever the reaction, Becker's approach is fascinating for the large number of predictions it generates about the institution. Consider divorce: Economists have learned from the theory of information that if one searches a market more thoroughly, one finds better prices at which to buy or sell. If one wishes a new car of given specification and accessories, one can save perhaps 1 to 2 percent of its price by going to three automobile dealers and buying from the cheapest rather than buying at the first dealership encountered. Similarly, Becker predicts that hasty marriages by the very young will be less efficient matches than those reached after extensive search (dating). Sure enough, the divorce rate is much higher for first marriages of people around age twenty than for those around age thirty, but the marriageability of people declines and divorce rates rise for marriages of people older than age thirty.

Again, consider polygyny, as practiced by the Mormons in the nineteenth century. The most successful men had several wives, the least successful often none. How did this practice affect the share of the family's income accruing to the women? Becker's answer is that it *raised* the women's share, simply because the supply of potential husbands had been increased, and that increased supply lowered the "price" a husband could command. Since the various wives' households were usually separate, the woman was the sole head of the family in residence much of the time. On this view, polygyny increased the share of family incomes going to women.

The work on the family, presented in 1982 in Becker's *A Treatise on the Family*, has had a large impact upon sociology, and it soon led to his appointment as a professor of sociology in that distinguished department at the University of Chicago. Becker has also made major contributions to labor economics and the theory of income distribution, so even the narrowest of economists must recognize his creativity. He may well go down in history as the chief architect in the designing of a truly general science of society.

Another area of large interest to economists is crime. After all, crime is an occupation, so it is a problem in labor economics: Who will become criminals, what crimes will they commit, and how much will they earn? Crime is also a major social concern: How can crime be contained and individuals persuaded to leave that calling?

The most definitive of crimes is murder, and it has received special attention from economists in recent times. The most influential economist has been Isaac Ehrlich, an able student of Gary Becker's. Ehrlich was the first person to make sophisticated analyses of the relationship between murder and capital punishment. Prior to his work, which first appeared in 1975, amateur statistical studies were interpreted to show that capital punishment had no effect upon murder rates, and it was argued that this was eminently reasonable. Murders were usually crimes of unthinking passion, and in addition, deterrence was provided as fully by long prison sentences as by capital punishment.

To an economist no truth is more firmly held than the one that as something gets more expensive, people buy less of it. ("Demand curves have a negative slope.") This, economists think, is surely true of potatoes, burglaries, even murders. And Ehrlich's studies confirmed this view. He found that when states executed murderers, for every person executed there was a substantial reduction in the number of murders—perhaps as many as eight to twenty less murders per execution. If Ehrlich is even remotely right—and later work adds support to his findings—a society must surely rethink its public policy. One can believe Ehrlich and still believe that for moral reasons the state should not execute people. However, one must then face the distressing implication that a policy of non-executions leads to sentencing a considerable number of anonymous individuals to death.

Ehrlich's work was also received by many economists and sociologists with the outrage we are now accustomed to observe: after all, in liberal circles capital punishment is viewed as both uncivilized and ineffective. The National Academy of Sciences appointed a committee to review the problem of criminal deterrence and in 1978 it produced *Deterrence and Incapacitation*. That committee did not deny the possibility of deterrence by capital punishment, but argued that the work to date was not sufficiently persuasive to support a change in public policy. The Academy enjoys the role of being the scientific arm of the federal government, and that enjoyment rests upon its leadership's careful attention to the winds of political opinion.

Of course, one should always seek to learn more about matters as grave as murder and execution, but the episode illustrates a trait that is widespread. People demand much higher standards of evidence for unpopular or unexpected findings than for comfortably familiar findings. This is not always a bad stance: The status quo usually has the endearing attribute that one usually knows pretty much what to expect from it—the average and the worst outcomes. Still, that is not an infallible rule. In particular, we know that during a period when capital punishment in America was virtually abolished (1960–1970), the annual number of murder victims rose from 8,464 to 16,848, or almost seven times as fast as the population grew. That status quo was a status low.

Several objections seem instinctively to arise in many minds when economic theory is applied to subjects like social discrimination, childbearing, marriage, and crime. The first objection is that all that is done is to restate familiar behavior in the language of a strange theory of behavior. That objection is clearly mistaken: When one can assert and even give rough measures of the deterrent effect of punishments upon crimes of passion as well as crimes for profit, we achieve a better understanding of crime and a handle to deal with it. When we can identify a potentiality for profit for entrepreneurs in circumventing discrimination against a group in the society, one may enlist a powerful force for the reduction of discrimination.

The second objection is that human behavior consists of much more than the rational selection of methods of achieving given ends subject to conditions imposed by nature or society. Is not a good deal of behavior altruistic rather than self-serving? Is there not much behavior that is incomprehensible on the theory that people are intelligently pursuing given goals? How can so simplified a view of human motivation be an adequate basis for explaining human behavior?

One answer to this second objection is that of course it is true; people also experiment, make errors, engage in irresponsible acts, are carried away by noble or absurd causes, and so on. None of these other forces is so powerful as the rational pursuit of goals, however; none is so strong, so persistent, so widely operative. There is good reason for this. The survival of the species requires that we pursue our interests with reasonable skill. When a motive other than simple self-interest appears to be important and widespread—an example is kinship altruism —it can be and is incorporated into the economist's theory. A second answer to the objection is that a remarkably wide range of the economist's predictions of human behavior proves to be correct. And that is answer enough.

The Imperial Science

Each time I have mentioned a new field that economists have invaded, it has been appropriate to say that they were unwelcome to the specialists in that field, whether politics, sociology, or law. That is perhaps to be expected: Who wants to be told to learn a new language and a new approach to familiar problems? It is more surprising that most economists, at least initially, have been unenthusiastic. Shouldn't they welcome an increased area of work and a larger demand for their services? The explanation, I believe, is the same for the economists as for the political scientists and sociologists: Older economists too are asked to work on unfamiliar problems, new kinds of data, unusual legal and social institutions. In short, the knowledge of all scholars became less complete and less current; they all became a little more obsolete. I have already referred to Max Planck's remark that a field progresses by having its old professors die off.

The prospect that economic logic may pervade the study of all branches of human behavior is as exciting as any development in the history of economics, or, for that matter, in the history of science. (Indeed, branches of the sociobiology literature find an extensive role for logic in animal behavior.) That does not mean that every social scientist need become an economist, any more than that every scholar who deals with inferences from observational data need become a statistician. Economics has been an exciting field for half a century, and I predict that this imperialism will reinforce that excitement in the decades ahead.

The Scope of Economics

Once upon a time economics was defined as the study of the production and distribution of wealth. How did it ever get involved in marriage, religion, politics, and even animal behavior and physiological structure? One way to answer the question is to observe that it is fashionable slang to say that economic logic is encapsulated in the phrase, "There's no such thing as a free lunch."

The slogan may have come from the sign in old-fashioned saloons, where a tray of sandwiches labelled "free lunch" encouraged patrons to drink most of their lunches. Of course, those lunches were charged for in the price of the drinks, and hence the slogan. But it goes deeper than that. Anything that is an object of conscious desire must be scarce: One does not consciously desire the air breathed, or to hear bad jokes. Scarce things are costly. If they weren't, everyone would get so much of each that they would not be scarce anymore. So anything scarce, and worth having, has been costly for someone to obtain. It wasn't free to them, and if they give it away, the chances are that the recepient will end up giving something valuable in exchange. Whether the lunch is paid for by giving the host an order for goods, or by inviting the host to be the guest next time, or by listening patiently to a bore, a price will be paid.

But what a way to define a science! It's rather like defining astronomy as the study of things that are usually more visible at night. Still, non-free lunches get to the essence of the formal definition of economics that is most widely employed: Economics is the study of purposive behavior involving choice. There needs to be a purpose, whether it be survival or warmth or companionship or avoidance of boredom—or a midday repast. There needs to be choice; choice of when to do something, how to do it, or with what means (resources) or with whom. Generously interpreted, that non-free lunch is the microcosm of economics.

CHAPTER 14

Ancestor Worship
and Abuse

Mᴀᴄᴀᴜʟᴀʏ (who is obviously a favorite of mine) once argued in characteristically vivid fashion that all dead people are good:

> These are the old friends who are never seen with new faces, who are the same in wealth and in poverty, in glory and in obscurity. With the dead there is no rivalry. In the dead there is no change. Plato is never sullen. Cervantes is never petulant. Demosthenes never comes unseasonably. Dante never stays too long. No difference of political opinion can alienate Cicero.[1]

Of course "good" here means "incapable of doing new

---

[1] Thomas B. Macaulay, *Critical, Historical and Miscellaneous Essays and Poems*, vol. 2 (New York: R. Worthington, 1879?), 144.

harm," but I believe that dead economists are good in a more positive way.

I have long been interested in how economists (and for that matter, the general class of scholars) behave and why. One might think that one's contemporaries are a better source of knowledge on this question than dead scholars, and in some ways they clearly are. One can argue with contemporaries and give them an opportunity to explain their ideas and acts. One can find out more about contemporaries than is easily learned about the dead: Are they stingy or generous, do they love their spouses, are their children monsters or gems, do they know a lot about subjects on which they do not write? One can of course dispute the usefulness of such knowledge, which seems more helpful in judging a scholar's character than in understanding his ideas.

Unless one is extraordinarily aggressive and socially skilful, one cannot cultivate many eminent scholars— and they are surely the most interesting to study. Indeed, friendship itself would seldom flourish in a relationship of close and objective study of a person; there is an implicit manipulative element in such a relationship. No such problem of mutual personal entanglement exists with the dead.

I have been interested in the intellectual history of economics since I wrote my doctor's thesis in this area, and that interest certainly was not the product of a conscious comparison of the benefits of studying the living or the dead. The history of economics is no longer a major academic subject; it is not taught at a professional level in most great universities, but the subject has not lost its charm for me.

One attribute of able economists that I learned early is that they seldom admit or correct a mistake. Of course an able economist shouldn't (and perhaps is not allowed to) make many mistakes, at least the kind of mistakes that are easily correctible, but even the few correctible mistakes are seldom admitted.

Consider the famous case of Adam Smith's defense of laws placing a maximum on interest rates (usury laws):

> The legal rate, it is to be observed, though it ought to be somewhat above, ought not to be much above the lowest market rate. If the legal rate of interest in Great Britain, for example, was fixed so high as eight or ten percent, the greater part of the money which was to be lent, would be lent to prodigals and projectors, who alone would be willing to give this high interest. Sober people, who will give for the use of money no more than a part of what they are likely to make by the use of it, would not venture into the competition. A great part of the capital of the country would thus be kept out of the hands which were most likely to make a profitable and advantageous use of it, and thrown into those which were most likely to waste and destroy it. Where the legal rate of interest, on the contrary, is fixed but a very little above the lowest market rate, sober people are universally preferred, as borrowers, to prodigals and projectors. The person who lends money gets nearly as much interest from the former as he dares to take from the latter, and his money is much safer in the hands of the one set of people, than in those of the other. A great part of the capital of the country is thus

thrown into the hands in which it is most likely to be employed with advantage.[2]

This is a strange argument; it seems to assume that lenders would pay no attention to the probability of being repaid, but only to the promised interest rate. Surely it was inconsistent with Smith's basic theory of sensible economic behavior; here the lenders are being foolishly shortsighted. The error was pointed out in a celebrated tract, *Letters on Usury* by Jeremy Bentham, but it must already have been questioned by others. Bentham's attack in 1787 came shortly before the final, trifling revisions Smith made in the fifth and last edition of *The Wealth of Nations* (1789), so his failure to retract the doctrine may be due only to lethargy—but also it may be due to his continued belief in the essential validity of his views.

A considerable list could be presented of large and small mistakes that have been made by economists (and a respectable list of my own mistakes), and seldom would it be necessary to add "but on this occasion the error was handsomely conceded." Sometimes, indeed, the author defends the error to the death, or possibly longer: A famous Viennese economist, Eugen von Böhm-Bawerk, would not admit the validity of the use of simultaneous equations; for him they involved circularity of reasoning.

Often the admission of error is surreptitious. Alfred Marshall asserted, in the first edition of his *Principles* (1890), that the total utility of a person's wealth is equal

[2] Adam Smith, *The Wealth of Nations*, vol. 1 (1776; reprint, Oxford: Clarendon Press, 1976), 357.

to the sum of the utilities derived from each commodity consumed. By the third edition in 1895, this passage was explicitly denied ("we cannot say that the total utility of the two [commodities] together is equal to the sum of the total utilities of each separately"), with a brazen footnote added: "Some ambiguous phrases in earlier editions appeared to have suggested to some readers the opposite opinion." (The problem arises in part because there would be double counting if two goods—say, beer and wine—served the same desires.)

Most mistakes are not of great consequence: Smith would not have had to make any changes in the rest of his big book if he had dropped the paragraph on usury laws. The reluctance to acknowledge mistakes is perhaps part of the committed missionary zeal of scientific explorers.

A second and related trait of scholars is that they seldom change their minds. We must remember that we are discussing scholars with good minds, and they would usually lose in a swap of minds. It was most unusual when Keynes's *General Theory* led Alvin Hansen, a hitherto conservative economist of mature years, to abandon his previous scientific position and become a vigorous exponent of Keynesian doctrine. The long disputations between Malthus and Ricardo from 1812 to 1823 are much more the pattern. These intelligent, honorable men, and warm friends, never seemed to reach agreement. A letter from Ricardo to Malthus a few days before the former's untimely death closed with grace:

And now my dear Malthus I have done. Like other disputants after much discussion we each retain our own opinions. These discussions however never influence our friendship; I should not like you more than I do if you agreed in opinion with me.[3]

The tenacity with which people hold the ideas in which they have a proprietary interest is not due simply to vanity. A scholar is an evangelist seeking to convert his learned brethren to the new enlightenment he is preaching. New ideas encounter formidable obstacles, the foremost being indifference, but also the new ideas will often conflict with old ideas or clash with apparently contradictory experience. A new idea proposed in a halfhearted and casual way is almost certainly consigned to oblivion. A scholar who cannot convince himself that what he proposes is certainly true and possibly important is asking a good deal of others to generate enthusiasm for the idea.

Another aspect of this salesmanship is the heavy use of repetition, perhaps the most powerful of arguments. I once served on a committee with John D. Black, the famous Harvard agricultural economist, and a master of the technique. He would state a position and I would offer one or two conclusive (to me) objections to it. His response would ignore my points and merely restate his position. I might then present one or two lesser objections to his position, chiefly to avoid boredom, only to founder again on his invincible obstinacy. Frank Knight, another master of repetition, once replied to this charge in the course of perhaps his ninth repetition of his views in an article on capital theory, by quoting Herbert Spencer, "Only by constant iteration can alien truths be

[3] *The Works and Correspondence of David Ricardo*, vol. 9, ed. by P. Sraffa (Cambridge, Eng.: Cambridge University Press, 1952), 382.

impressed upon reluctant minds." The method serves equally well with alien errors.

Scholars write to influence the beliefs of fellow scholars, but "influence" is itself an ambiguous idea. To influence can mean to persuade others to buy one's book; to read one's book; to cite one's publications; to believe differently than before, but not necessarily what the persuader desires; to accept the writer's arguments; and so on.

Some kinds of influence can be measured with tolerable accuracy. One can count the sales of a book, and estimate the probable number of readers by surveys of reading practices or the number of times the book or article is cited. Yet these measures do not tell me whether the work influenced the course of professional thought and work. I may cite with approval Becker's theory that criminals behave rationally in their response to prospective punishments, but does that mean I got the idea from him or that he sharpened my understanding of the theory or am I merely using him to reinforce the plausibility of a view I already held?

On rare occasions the extent to which a given influence affects beliefs is easy to identify. A late friend, G. Warren Nutter, was a fine economist at the University of Virginia. He once embarked on an airplane to go to the University of Rochester to give a seminar on why the famous Coase Theorem was wrong. Milton Friedman happened to sit next to him on the first leg of the journey, and they discussed Nutter's argument against Coase. When Nutter arrived at Rochester, his lecture had changed to something like "Another Way to Prove the Coase Theorem."

The most skeptical observer cannot cavil over the source of influence here.

Most influence is not so simple as when A converts B to A's view of things. Much more often B seeks to refute A, and surely this represents an important influence of A. When Keynes produced a bundle of paradoxes and heresies in his *General Theory of Employment, Interest, and Money* in 1936, a fair fraction of leading economists denounced or defended him. For example, he stated that every dollar that is saved by a society is invested by that society: Savings and investment are always equal. This proposition was a tautology with Keynes's definition of the terms, but a lot of literature was generated in the process of establishing this fact. Again, the later attacks by Friedman on Keynes for underplaying the influence of changes in the stock of money on economic activity called forth extensive defenses of Keynes by James Tobin, Franco Modigliani, and others. It is an achievement when others think that one's arguments are important enough to be denounced and demolished. Even to be demolished is better for one's self-esteem and reputation than to be ignored: It requires some ability to excite and especially to outrage one's fellow professionals.

Some of these traits of intellectual leaders are caught in the statement that they lack a sense of humor. I mean by this, not the inability to laugh at the right point when hearing a joke, but the ability to view oneself with detached candor. Ridicule is a common weapon of attack but amused self-examination is a form of disarmament; one so endowed cannot declaim his beliefs with massive

certainty and view opposing opinions as error unconta-
minated by truth.

Scholars have a strong penchant for referring to their
intellectual ancestors, whether favorably or unfavorably,
and usually whether they have read the works of these
ancestors or not. One historian of economics, D. H. Mac-
Gregor, said that Say's Law (a proposition, not really due
to J. B. Say, only partially summarized by saying that
there cannot be overproduction of all commodities at one
time) should be called Hearsay's Law. Jacob Viner, whose
vast and honest erudition has long been my despair, once
told me that the average modern reference to the classical
economists is so vulgarly ignorant as not to deserve notice, let alone refutation. I shall not give examples, but
famous economists have made breathtaking misrepresen-
tations of Malthus on population, Ricardo on value, and
so on.

This practice raises the question: Why do people make
statements about earlier scientific work that are simply
ignorant? Is it to convey a false impression of historical
learning, or to emphasize the wide historical perspective
that underlies contemporary ideas? Or is it simply to label
oneself subtly as a follower of or rebel against some earlier intellectual tradition? I suspect that all of these explanations have some role. They support Bishop Stubb's
famous remark that when someone says "History
proves," that phrase should be replaced by "I propose to
assume without a shred of evidence."

And this brings us to the fundamental question: Why
should a scientist be acquainted with the history of his
subject? It is nice to be widely educated, no doubt, but
how does it help a modern economist who is working on

oligopoly theory or business cycles to study the economic literature of the Victorian period? Is not everything that is valuable in earlier thought refined or distilled into the modern treatises and textbooks?

The answer given by modern scholars is an emphatic affirmative: Yes, (almost) all that is valuable, and some that is not, is incorporated in the received knowledge. Just as a student of differential calculus would learn little if he turned to Newton or Leibnitz or Lagrange, so a student of price theory is unlikely to learn from Adam Smith or, in a work a century later, from Alfred Marshall. When I say that this is the prevailing answer, I mean both that it is the practice of modern economists and that the practice does not interfere with brilliant scholarly careers. A young economist who believes that Adam Smith was the Smith who founded the Mormon faith will only provide innocent amusement outside of Utah, at no cost to his professional status.

It is always difficult to win a quarrel with success, and I shall not try. In fact, I cannot be confident that it would be profitable for a young scholar to study the history of his subject. If a young economist does immerse himself in the history of economics, he will learn that every proposed innovation is first produced in a highly imperfect form, and only gradually will the larger imperfections be removed. He will also learn that the sponsors of the new theory or program exaggerate its merits and just as consistently exaggerate the deficiencies in the previous knowledge they are seeking to displace. I know of no important exceptions to this pattern of aggressive academic salesmanship.

For example, Adam Smith carefully ignored the interesting treatise on economics by Sir James Steuart that had appeared nine years before. (Neglect is the highway to oblivion.) The one conspicuous exception to the rule of overestimation of the importance of one's own ideas was John Stuart Mill, whose rectitude was so extreme as to be painful: *He* played down his own contributions—and was rewarded for a century with an undeserved reputation for noncreativity. So modesty and respect for received knowledge would be most dubious assets for a scientific innovator. This lesson, on reflection, should be obvious enough; one cannot communicate effectively with other people unless one uses the language to which they are accustomed.

That is one side of the ledger, and there is another side. The writings of economists long departed allow a degree of detachment for a reader that is probably impossible to achieve with living writers. One surprising feature taught by intellectual history is the persistence of uncertainty over what a person really meant. One might think that intellectual competence and goodwill are all that are required to understand what a scholar intends to say, but the study of any important scholar of the past will show that belief to be most naive. One can produce 170 years of disputes, for example, over how Ricardo believed that wage rates were determined. Did he believe that wage rates of unskilled labor could for long periods exceed the cost of subsistence, or did the potentials for rapid population growth soon bring wages back to a subsistence level? Was the subsistence level a biological minimum or some culturally determined conventional standard? If the latter, was the level stable over time? One reason these disputes persist is that Ricardo's text is often ambiguous: Page X takes or implies one position and page Y another. That sort of ambiguity is not due simply to carelessness, for at one point he may have been thinking of the short run and at another of the long run, or at one point the focus is on another topic so the wage question is simplified to get it out of the way. Still, carelessness, or at least imprecision, is present in significant measure in the writings of all scholars.

Another source of ambiguity is the dependence of the meaning of an idea on one's knowledge. When an economist in the time of John Stuart Mill (circa 1848) wrote that there is only one price for a commodity in a competitive market, he was reproducing a widely accepted approximate truth (although Mill called it an axiom). The argument is given in chapter 5: Informed buyers would seek the lowest price offer, and informed sellers would seek the highest bid, so differences in prices actually paid would be eliminated. Today, however, it is no longer permissible to assume tacitly that buyers and sellers are fully informed; information is never perfectly free (it costs a newspaper to learn the prices of common stocks). We now believe that one will find a variety or distribution of prices in almost all markets, with the extent of the variety depending largely on the cost of information.[4]

---

[4] One advantage of the study of the history of a science is that it informs one of the value of books. My son Stephen found a splendid copy of Heinrich Gossen's famous, extraordinarly scarce 1854 book on utility theory and bought it for $25. He gave it to me as a Christmas present. Ten years later another copy went for $15,000 at an auction. I claim a little credit: Stephen is a learned historian of statistics but would he have known of Gossen without an economist father?

Mill knew this too, but he recognized the variety of prices only occasionally and incompletely. What was standard practice in 1848 is ambiguity today.

A second useful and somewhat surprising lesson of historical scholarship is that widely accepted facts are often wrong. Not minor facts suitable for a game of trivia, but pervasive facts that have a powerful influence on how a science thinks and works. I shall give two examples. Most English economists thought that a landowner need do nothing but rent out his land: All he had to sell was access to the soil. Hence they usually recited that "the landlord loves to reap where he has not sown," and questioned the ethical basis of rent payments. Of course this was absurd. The skill of the landowner in making investments and in determining when to change the use of his land had an important influence on the yield of the land. From Adam Smith through Henry George and George Bernard Shaw this fact was consistently ignored.

Again, the belief that wages of unskilled workers were approximately stable over long periods was widely accepted in the first half of the nineteenth century. To his dying day (May 8, 1873), John Stuart Mill believed that population pressure was the fundamental barrier to the improvement of the living conditions of English (indeed, all) workers. No serious test of the "fact" was made until Robert Giffen began his studies in 1875 of the incomes of working men, and of course it proved to be false. Real wages had risen by 50 to 100 percent in the previous half century.

Much more modern examples could be given. For example, the belief that most American manufacturing industries are monopolized is wrong, but it was long treated as approximately true. It is tempting to scold the economists for not making perfectly feasible studies to determine the facts, but it has happened too often to be attributable to indolence or special historical conditions. In each such case the "fact" has had some empirical basis. Once it became widely accepted (by a process no one has studied), it is likely to persist until it becomes untenably anachronistic (as with Malthus's population theory) or it is challenged by theorists for whom the "fact" is troublesome.

So one valuable service of the past is to teach us how limited the knowledge and imagination of even the finest scholars are. We all wear glasses that carry a date in time and the name of some geographic area, and with even the keenest of vision these glasses allow us to see only limited distances and partial motion of our world.

If that is a sobering lesson, then there is a special advantage of the past: It allows the association with superior minds. No university will ever have at one time four economists of the quality of Adam Smith, David Ricardo, Irving Fisher, and Alfred Marshall, to say nothing of a dozen of their best colleagues—but they can all reside in one's library. Their subtle minds are ever ready to instruct and tease and baffle. They teach that they become incomprehensible if they are read with a high-powered microscope, and are hopelessly bland if read with a telescope. One can marvel at how they succumb to their personality and environment at one point, and utterly ignore them at another. A superior mind and its products must be the most fascinating of scholarly objects—and here they are, available at the cost only of one's own intellectual effort. I have mentioned the economics of free lunches: Here is a long sequence of marvelous repasts, all the more marvelous because they improve as our understanding improves.

INDEX

Index

223

Index

Index